OPEC

OPEC

The Failing Giant

MOHAMMED E. AHRARI

THE UNIVERSITY PRESS OF KENTUCKY

Scholarly publisher for the Commonwealth,
serving Bellarmine College, Berea College, Centre
College of Kentucky, Eastern Kentucky University,
The Filson Club, Georgetown College, Kentucky
Historical Society, Kentucky State University,
Morehead State University, Murray State University,
Northern Kentucky University, Translyvania University,
University of Kentucky, University of Louisville,
and Western Kentucky University.

Editorial and Sales Offices: Lexington, Kentucky 40506-0024

Library of Congress Cataloging in Publication Data

Ahrari, Mohammed E.
OPEC : the failing giant.

Bibliography: p.
Includes index.
1. Organization of Petroleum Exporting Countries.
2. Petroleum industry and trade. I. Title.
II. Title: O.P.E.C.
HD9560.1.O66A64 1986 341.7′5472282′0601 85-15040
ISBN 0-8131-1552-3

To my father and in remembrance of my beloved mother

عمر بھر تیری محبّت میری خدمت گر رہی
میں تیری خدمت کے قابل جب ہوا تو چل بسی

"Your love for me sustained me ever so long; but when I attained the position to render service to you, I was deprived of the privilege by your death."

Rough translation from Mohammed Iqbal's poem, "In Remembrance of My Mother"

And to Rheana with love

Contents

Tables and Figures

Tables

Appendix

Figures

Map

Abbreviations Used

AIOC: Anglo-Iranian Oil Company
API: American Petroleum Institute
APOC: Anglo-Persian Oil Company
BNOC: British National Oil Company
BP: British Petroleum
CFP: Compagnie Française de Petrole
CIA: Central Intelligence Agency
CIEC: Conference on International Economic Cooperation
CPI: Consumer Price Index
ECG: Energy Coordination Group
EEC: European Economic Community
ENI: Ente Nazionali Idrocarburi
IEA: International Energy Agency
IEP: International Energy Program
IMF: International Monetary Fund
INOC: Iraq National Oil Company
IPC: Iraq Petroleum Company
KOC: Kuwait Oil Company
Linoco: Libyan National Oil Company
NIEO: New International Economic Order
NIOC: National Iranian Oil Company
NOC: national oil company (of OPEC)
OAPEC: Organization of Arab Petroleum Exporting Countries
OECD: Organization for Economic Cooperation and Development
OPEC: Organization of Petroleum Exporting Countries
Socal: Standard Oil Company of California
Socony: Standard Oil Company of New York
Texaco: Texas Oil Company

Acknowledgments

A project as comprehensive as this one involves many individuals other than the author. I owe a special debt of gratitude to my wife, Rheana, not only for the lay perspective she contributed early in the work but also for her patience, moral support, and love throughout the writing process. I also wish to thank Professor Charles T. Goodsell of the Center for Public Administration and Policy at the Virginia Polytechnic Institute and State University for reading the first draft and making valuable suggestions; Professor Jo Ann Jones of the Department of English at East Carolina University for her meticulous editorial assistance; Ms. Patricia R. Guyette, and Ms. Artemis Kares and Mr. Michael C. Cotter of the Joyner Library at East Carolina University for helping me during the arduous research that went into this study; Mrs. Cynthia M. Smith of the Political Science Department for promptly and cheerfully typing draft after draft; and the two anonymous reviewers for the University Press of Kentucky for their critical insights. Finally, I wish to acknowledge my parents for dedicating their lives to eliminating uncertainties during my formative years. The dedication of this book to them is only a minor token of my utmost love, respect, and gratitude.

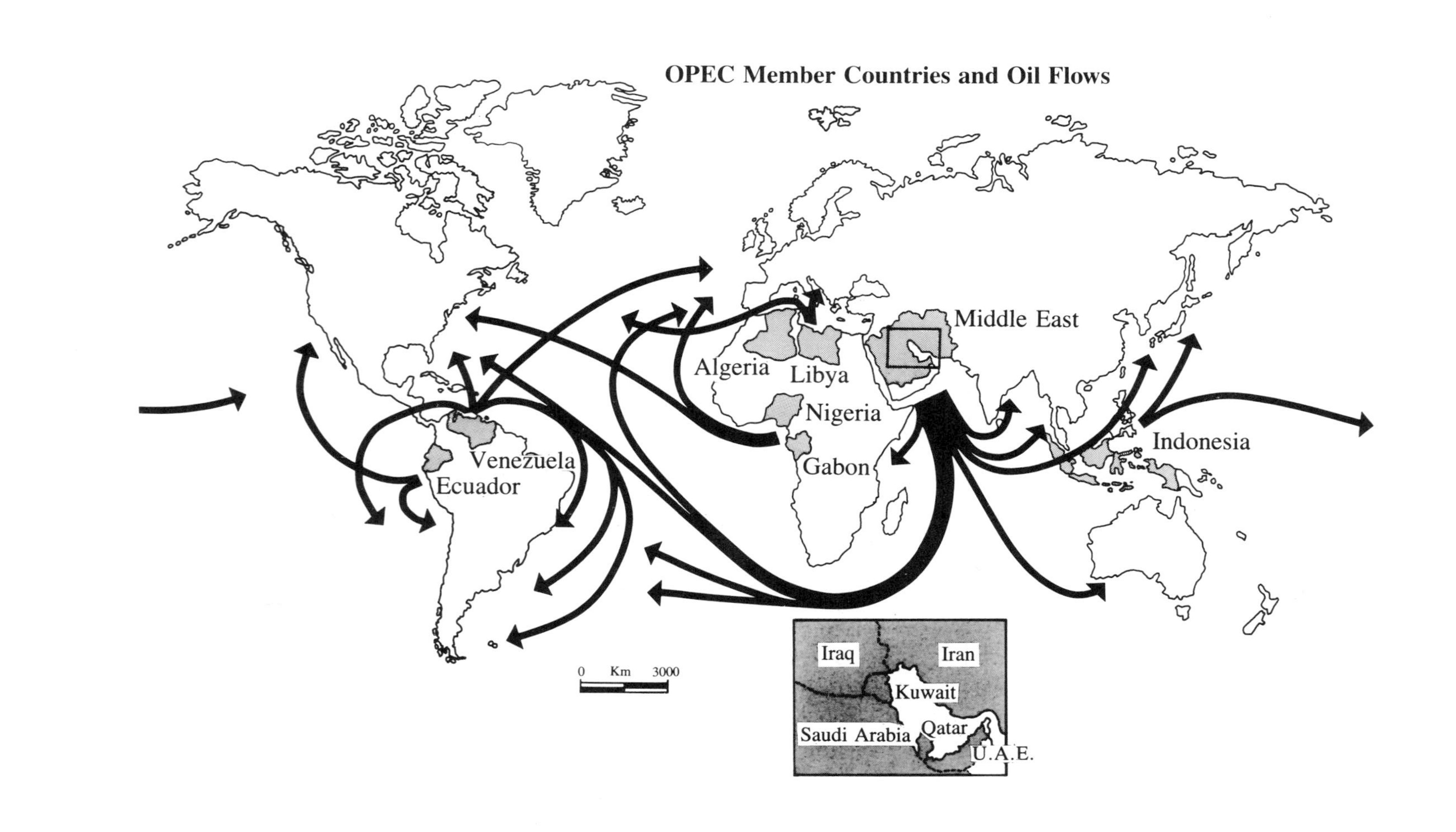
OPEC Member Countries and Oil Flows
Middle East
Algeria
Libya
Nigeria
Gabon
Venezuela
Ecuador
Indonesia
0 Km 3000
Iraq
Iran
Kuwait
Saudi Arabia
Qatar
U.A.E.

1. Introduction

The oil affairs of the non-Communist world went through significant mutations between the 1960s and the early 1980s. In sharp distinction to the 1970s, the 1960s were marked by an abundance of cheap oil. Ample supplies of oil favored the consuming nations and discriminated against the oil states, which were struggling, with little success, to raise the price of their commodity. Moreover, the supply disruptions that were to come had not been seriously anticipated by decisionmakers in the Western industrialized countries. In stark contrast, throughout the 1970s the international economy suffered from a variety of symptoms of the "oil crisis," which caused considerable consternation among the oil-consuming nations. The early 1980s, on the other hand, were marked by the prevalence of an "oil glut" that threatened the economic well-being of the petroleum-exporting countries. These markedly contrasting conditions, appearing during a period of about twenty-five years, are referred to in this study as the uncertainties of the buyers' and sellers' markets.

This book examines the nature and dynamics of those uncertainties. The paradox of oil power arises from the ability of a group of advantaged actors—the oil industry in the 1960s and the petroleum-exporting countries in the 1970s—to bring about changes that are most beneficial to themselves and at the same time to minimize the deleterious effects of their exercise of oil power. The paradoxical nature of oil power is much more apparent for the oil states than for the multinational oil corporations or the industrial consuming countries.

The core of this study is an examination of the Organization of Petroleum Exporting Countries (OPEC) and its performance as an economic alliance from 1960, when it was founded, through the first quarter of 1985, when the continuance of a soft market and the potential for continuation of similar conditions in the years ahead rendered its future cloudy. The non-oil-producing developing countries are peripherally mentioned here, but Communist countries are excluded. The main thesis of this study is that the pricing behavior of OPEC is essentially economic in nature. It is argued in these pages that, even though one cannot totally exclude the dimension of politics from OPEC's decisions, the overriding factor governing these decisions was the desire either to maximize economic payoffs or to minimize the harmful effects of such phenomena of the international economy as inflation, devaluations of the dollar, and recession. Toward this end, an analysis is made of important meetings of OPEC between 1960 and early 1985. Even in light of the Arab oil embargo, which was imposed by the Organization of Arab Petroleum Exporting Countries (OAPEC) and was essentially politically motivated, the argument is advanced that the decision to proceed with this action was made in an environment highly advantageous to the economic objectives of its participants. In other words, the oil embargo had no negative effect on the revenues of the Arab oil states.

As a starting point, it is essential to elaborate on the meaning of "uncertainty," which is used here as an aggregate term, and to spell out as precisely as possible what the various actors were uncertain about. The uncertainties of the sellers' or the buyers' markets involved questions of supply and price of crude oil and their effect on producers, consumers, the oil companies, and the international economy. The buyers' market of the 1960s had the following characteristics: (1) the buyers of crude oil—the multinational oil corporations and oil consuming nations—had a clear advantage over the sellers. (2) In general, the supplies of crude oil outlasted prevailing demands. (3) The prices of crude oil were depressed, and long-range prospects of price increases were not optimistic from the vantage point of the sellers.

In this buyers' market the oil states remained uncertain of their own ability to bring about what they regarded as "fair" increases in the prices of crude oil. Such uncertainty had far-reaching implications for the oil states, since it was inextricably related to their ability to finance economic development and to promote social tranquility in their polities. In fact, the very chances of their regimes' surviving were envisaged by the oil states as being related to their ability to extract increased oil revenues from foreign concessionaires. Such was the plight of the oil-producing countries during the 1960s.

The sellers' market of the 1970s had the following characteristics: (1) The petroleum-exporting countries had a clearcut advantage over the consuming nations and the oil industry. (2) The market was characterized by recurring crude oil shortfalls, and even the reemergence of a soft market was no guarantee that such a condition might not be abruptly replaced by a tight market. (3) The prices of crude oil were raised periodically, and the long-range prospects of such escalations appeared promising for the producing nations.

The oil market of the early 1980s was characterized by low demand for OPEC oil because of a combination of the general economic downturn in the industrialized consuming nations, the sustained practice of conservation and fuel-switching, and increased competition from non-OPEC petroleum-exporting countries.

The Analytic Approach

Price- and supply-related uncertainties are endemic to oil affairs. The establishment of OPEC was an institutional response by the oil-producing states to two unilateral decreases in the price of crude oil, the first in February 1959, the second in August 1960. This organization was created to restore the price of oil to the pre-August 1960 level and to present a united front against the powerful oil companies; without such a front, the oil states were convinced they could not prevent the companies from unilaterally cutting oil prices. Throughout the 1960s, nevertheless, OPEC continued to be the victim of the depressed prices of the buyers' market then prevailing.

Toward the end of the 1960s, a variety of events and factors led to the emergence of a sellers' market and its attendant uncertainties for consumers, which included intermittent leapfrogging in the price of oil by the oil states as a result of a series of price negotiations between 1970 and 1973 and thereafter. By and large, the OPEC members fully exploited the sellers' market to correct what they perceived to be an age-old exploitative relationship between themselves and the multinational oil corporations and their parent countries. In this sellers' market the oil industry remained uncertain about the longevity of the negotiated settlements because the oil states continued to renegotiate previous agreements. From Tripoli I, the first negotiated agreement, to the participation agreements, the oil states negotiated a total of six agreements, five of which could be characterized as renegotiations. The fourth quarter of 1973 marked the beginning of an era in which the oil industry and Western consuming countries became uncertain not only about prices but also about the availability of oil.

The sellers' market produced mixed results for the oil companies. The continued shortages of supplies, created by the manipulations of OPEC members, and periodic increases in the prices of crude oil enabled the oil states to conclude the participation agreements expeditiously. These agreements resulted in a *de facto* nationalization of oil. The oil companies begrudgingly lost their status as concessionaires[1] and were forced to play the role of managers in the upstream operation. But the continuation of the sellers' market brought about little change in the downstream operation of the oil companies. In fact, their continued predominance in this type of operation remained the single most important reason for their sustained ability to reap a bonanza even during the peak hours of the sellers' market.

The prolonged existence of the sellers' market, however, left an indelible imprint on the international economic system in the form of inflation, unemployment, and recession. The industrialized consuming nations were plagued by stagflation,[2] while the economies of the developing countries were crippled by lack of growth and acute inflation.[3] For the Western industrialized countries, the sellers' market caused a considerable amount of alarm, especially because cheap and secure supplies of oil had become a *sine qua non* of their way of life for several generations.

Despite the power and prestige acquired by the oil states in the 1970s, they were unable to incorporate long-term strategies that would stabilize the prices of crude oil and would insulate their economies from the erosive impact of international inflation. And despite their attempts to keep pace with inflation through raising the prices of their commodity in a series of agreements made between 1971 and 1973 (Tripoli I, Teheran, Tripoli II, and Geneva I and II) inflation rates remained mostly ahead of them. Toward the end of 1973, however, OPEC did succeed on two occasions in introducing quantum increases in the prices of oil. The first escalation occurred in October, immediately following imposition of the Arab oil embargo, and the second took place in December, as the non-Communist economies were experiencing supply shortages resulting from the embargo. Without cooperation between the oil-consuming and oil-producing nations, however, inflation remained uncontrollable. OPEC's 1973 price increases contributed substantially to the already rampant inflationary trends; as a result, the non-Communist countries experienced a recessionary period in 1974 and 1975.[4] Under the continuing uncertainties of the sellers' market, the economies of all nations seemed to be affected in varying degrees by international inflation, unemployment, and recession. The oil states, however, continued to fare better than the non-oil-producing countries.[5]

Even though OPEC members played no deliberate role in creating the supply shortages that lasted from 1979 through the first half of 1981, they fully exploited these conditions by introducing a series of unrealistic price increases. In contrast, under soft market conditions in the period 1975-1978 and again after 1981, the oil-producing nations seemed to drift with the ebb and flow of events, waiting for another political crisis to tighten the market for them.

The 1980s appear to be a significantly different decade from the 1970s for OPEC members. At least in the first five years of the 1980s, the demand for OPEC oil remained depressed (see Table A-1 in the Appendix) because of a combination of continued recessionary trends in the industrial countries, conservation, fuel-switching, and competition from non-OPEC oil producers.

Bargaining in Oil Affairs

In oil affairs the notion of bargaining has a special meaning which needs elaboration. Bargaining between oil-producing countries and oil companies, between oil-consuming and oil-producing nations, and even among the oil states is subservient to the dynamics of the marketplace. For instance, the uncertainties of the buyers' market in the 1960s swung the advantage so excessively in favor of the oil companies and, by extension, of the consuming states (making them the advantaged actors) that the oil companies had no compelling reason to offer concessions to the oil states (the disadvantaged actors) in their quest for stabilization of crude oil prices or increases in royalties.

Similarly, in the 1970s, when the oil states became the advantaged actors, they proceeded to correct what they perceived to be gross and age-old injustices inflicted upon them by the multinational oil companies and by the structural inequalities of the international economic system. They introduced intermittent increases in the prices of crude as a direct response to their advantageous bargaining position. The eroding bargaining position of the oil companies forced them to approve almost every demand put forth by the oil states in 1971 through 1973, even before a new round of negotiations was initiated. Finally, in October 1973, when the convergence between supply shortages and periodic price escalations improved the bargaining position of the oil states so decisively that even a semblance of negotiation with the oil companies was deemed unnecessary, the oil states as a group emerged as the sole determiner of crude oil prices.

Starting in 1974, the oil states, in alliance with the non-oil-producing countries of the Third World, attempted to negotiate the establishment of the New International Economic Order (NIEO) with the Western industrial countries. Frequent negotiating attempts failed to produce a framework of cooperation between the two groups of actors, however, because the industrial nations deemed the NIEO-related demands unrealistic and because OPEC and its Third World allies felt that the other side was intransigent.

The unrealistic nature of these advantaged actors' demands was reflected also within OPEC membership. In the 1970s, Saudi Arabia, acting as a swing producer, resolved to impose its will on the price hawks (Iran, Libya, and Algeria, whose ranks were frequently joined by Iraq, Nigeria, Venezuela, and Indonesia) by escalating production rates and maintaining a lower price, thereby glutting and softening the market. The price hawks considered this action blatantly unfair. They defied it in 1977 but succumbed to it in 1981.

A final word is necessary on the nature of bargaining in oil affairs. Because the demands imposed by advantaged actors on disadvantaged ones were perceived by the latter as unrealistic, illegitimate, and unjust, most of the gains made by the advantaged actors can be considered temporary, for they are likely to be lost under market conditions which reverse the positions of strength. The prices of crude oil top the list of such unrealistic or unjust, but temporary, gains made by the advantaged actors. The participation agreements were gains of a permanent nature, however, and are not likely to unravel even if the oil market becomes a buyers' market once again. The present association between the oil states and the oil industry, wherein the oil states have sufficient technical knowledge to run their upstream operations and retain the multinational oil corporations only as service contractors and managers, is permanent enough to assure that there will not be a reversion to the concessionary arrangements that prevailed into the early 1970s. In the 1980s, a number of oil states have also made significant strides of a permanent nature in expanding their activities in refining of crude oil, shipping of oil, and even in participating in exploration for oil outside the OPEC area.

The Oil Companies and the Host Countries

As early as the late nineteenth century, the multinational oil companies, with the active support of their governments, initiated exploration in various regions of the world to satisfy rising demand and to find new sources of oil as production of petroleum declined in the industrial countries. As a result of this massive search for oil, British companies obtained an important concession in Iran, Dutch companies obtained a few concessions in the East Indies, German, British, and Dutch companies acquired a concession in the Ottoman Empire, and American and Anglo-Dutch companies concluded concession agreements in Romania, Russia, and Latin America.[6]

In the United States, fear of the early exhaustion of oil led to determined efforts by American oil corporations, with government backing, to conclude agreements similar to those negotiated by the European companies in areas such as the Middle East, the East Indies, and Latin America. The general outcome of this international scramble for oil was the emergence of eight major companies—the international majors—which formed an oligopoly of the oil industry. Five of these were American: Standard Oil of New Jersey (Exxon), Mobil Oil (Mobil), Texas Oil Company (Texaco), Gulf Oil (Gulf), and Standard Oil of California (Socal); one was British: British Petroleum (BP); one was Anglo-Dutch: Royal Dutch/Shell; and one was French: Compagnie Française de Petroles (CFP). By 1970 these companies controlled 80 percent of the volume of world oil trade.[7]

In addition, there was another group of twenty to thirty independent oil companies, referred to as the international minors because they were new arrivals and held a minor status in the market. A majority of these international minors were American firms, such as Standard Oil of Indiana, Phillips Petroleum, Continental Oil, Atlanta Refining Company, Union Oil, and others. Some important non-American international minors were the Japanese-Arabian Oil Company and the Ente Nationali Idrocarburi (ENI) of Italy.[8]

In general, the establishment of international oil companies as business entities in the various oil-producing countries took place at a time when the national governments in these countries existed in their most primitive form and when the level of social consciousness and political awareness of their peoples was also very low. In most instances, the colonial power determined the scope and nature of oil contracts (or concessions) with their oil companies. The local rulers, as a result of a combination of such factors as political ineptitude, sheer ignorance about the technicalities of the oil business, and fear of the consequences if they refused to cooperate with their colonial masters, rubber-stamped these contracts. Two examples provided by George Stocking, a specialist in the international oil business, are sufficient to establish these observations. Describing the political, social, and economic environment under which the D'Arcy concession, the first major concession of this century, was signed by the imperial regime of Persia (Iran), Stocking writes:

> On the assassination of his father in 1896 Shah Muzaffar ed-Din began a reign which was to witness the dissipation of the State's revenue, an increase in burdensome foreign indebtedness, and the alienation of the country's most important source of national wealth, oil. Accustomed to extravagant living at home and expensive travel abroad, surrounded by a corrupt and pretentious court, faced with an empty treasury, lacking in appreciation of the economic potentialities of Iran's oil resources, confronted by concessionseekers willing to pay for the privilege of risking their capital in a search for petroleum, and responsible to no one for any mistakes that he might make, the Shah signed the D'Arcy concession.[9]

Elaborating on the lack of political or technical sophistication of the Saudi ruler who signed the oil agreement with the Western oil companies, Stocking notes:

> An absolute monarch, unhampered by any of the paraphernalia of modern governments, restrained only by Moslem customs and the advice of his tribal and religious leaders, making no distinction between the public and the privy purse, badly in need of funds with which to meet the increasing cost of governing his loosely knit kingdom, unschooled in the ways of the West and ignorant of its technology, bargained as an Arab trader with sophisticated representatives of a modern oil company anxious to obtain a source of oil with which to supply its domestic needs and permit it to enter world markets.[10]

Given the gross political and technical asymmetry between the host governments (i.e., the oil-producing countries) and the international corporations, the oil concessions signed between them were bound to favor the latter lopsidedly. An oil concession, the sole legal basis for the host country-oil company relationship, consisted of a written agreement whereby the latter was granted rights by the former to explore, develop, and export oil from specified areas during a specified period of time. The host, in return for all these guarantees, received financial payments and other benefits such as royalties (i.e., the owner's share of profit) and taxes.

The D'Arcy concession, signed on May 29, 1901, was a classic example of lopsidedness. This concession, which was to last sixty years and covered 500,000 square miles of Persian territory, provided exclusive privileges ranging from exploration to exportation of natural gas, petroleum, asphalt, and ozocerite. The agreement exempted all lands and imported equipment from all taxation through the duration of the concession. The Persian government guaranteed security for personnel and property of the D'Arcy company. In return, the company agreed to pay the government within a month of its formation £20,000 in cash, £20,000 in stock, 16 percent of the annual profit, and a sum of $1,800 per year. Disputes between the government and the company involving the provisions of the concession were to be referred to two arbitrators, one of which was to be appointed by each party, and an umpire selected by the arbitrators. The decision of the arbitrators, or (in case they disagreed) that of the umpire, was to be final.[11] The significance of the D'Arcy agreement as the basic framework of the relationship between a host government and an oil corporation becomes apparent in the fact that all ensuing concessions provided the concessionaire companies with: (1) exclusive right to explore, extract, and trade; (2) huge territories of the host countries, in some instances almost the whole country (as in the cases of Iraq, Qatar, and Kuwait); (3) long duration (between forty-five and seventy-five years); and (4) exemption from all direct and indirect taxation.[12]

Aside from being lopsidedly in favor of the international oil corporations, the oil concessions were also highly lucrative for these companies and their mother countries. For instance, between 1911—when the Anglo-Persian Oil Company (APOC) had its start in the business—and 1951, the British treasury received $700 million in oil revenues, while Iran earned only $316 million. These figures do not include the enormous profits made by APOC (which later changed its name to Anglo-Iranian Oil Company—AIOC).[13] For 1950, one year before Iran nationalized its oil industry, Britain and Iran received $140 and $45 million, respectively.[14]

The oil states were becoming increasingly aware of the unjust nature of these concessions as far back as the early 1930s. Reza Khan, who came to power as Shah Pahlevi only seven years earlier, cancelled the D'Arcy concession in 1932 when the royalty payments to the Persian government fell from £1,288,000 to £306,000 a year. Harvey O'Connor, who has written extensively on oil affairs, points out that this cancellation was based upon several complaints Persia had against APOC: the government did not have access to the books kept by APOC in order to check the validity of payments; the government did not receive any earnings from the company's "lucrative" tanker fleet; the company was reporting less income than it was actually earning by offering special discounts to its subsidiaries; while the British government was receiving income from APOC, the government in Teheran received no such payments; and the company was not fully developing the oil potential of Persia.[15] The resentment stemming from this grossly asymmetrical relationship between AIOC and the Iranian government continued to simmer.

Toward the end of the 1940s, the 50-50 profit-sharing system was instituted in Venezuela. At the same time, the government of Venezuela was also becoming concerned about substantial production increases in the Middle East, where the cost of production was low, at the expense of lowered production within its own borders. Part of the reason underlying this escalation of production, as Venezuela perceived it, was the prevailing political and economic climate of the Middle Eastern countries, where the international oil companies enjoyed minimal financial obligations and encountered few political obstacles. In order to increase the political and economic awareness of the Middle Eastern governments concerning the oil business, Venezuela in 1949 sent a number of missions to these countries. Venezuela hoped that, once these governments became aware of the 50-50 profit-sharing system, the inordinate competitive advantage enjoyed by Middle Eastern oil would be lessened, thereby easing the downward pressure on Venezuelan production. As a result of these contacts, Saudi Arabia concluded a 50-50 arrangement with its concessionaire companies.

The introduction of a 50-50 profit-sharing regime in Saudi Arabia intensified pressure for a similar agreement in Iran. Toward the end of 1950, a special oil committee of the Iranian Majlis (parliament), chaired by Dr. Mohammad Mossadeq, initiated a study of nationalization. Assessing the winds of change, AIOC concluded an agreement with the prime minister, Ali Razmara, the details of which Razmara refused to disclose.[16] The prime minister waged an intense campaign against the rising demand for nationalization, but was assassinated on March 7, 1951, and on March 15 the Majlis voted for nationalization. Within one month after this action, Dr. Mossadeq became prime minister. Shah Mohammed Reza Pahlevi, the young son of the deposed Reza Shah, was forced to give his assent to the parliamentary action by May 1, and the measure became law.

For Britain, the nationalization episode in Iran brought a shocking awareness of its diminishing power in the post-World War II world. For AIOC, this action was a clear signal that the monopoly of multinational corporations might not last much longer. Neither Britain nor AIOC was ready to roll over and play dead, however. Recognizing the newly emerging world power relationship, Britain immediately ruled out military action to nullify nationalization. Other courses of action, such as the use of the International Court, economic boycott, diplomatic pressure, or even covert actions aimed at overthrowing the government of Dr. Mossadeq, were still available to Britain, however. The refusal of the International Court to accept jurisdiction in this case put an end to this course. The British government and AIOC orchestrated the two-pronged strategy of instituting an economic boycott of Iranian oil and applying diplomatic pressure on other governments to refrain from purchasing petroleum products from Iran. AIOC and Britain compensated for the loss of oil revenues from Iran by increasing production in the oil fields of Kuwait.

These actions did not, however, bring about the collapse of Mossadeq's government, as anticipated by Britain and AIOC. Because of the insubstantial amount of oil revenues Iran was accustomed of receiving from its concessionaire, the Iranian economy was not heavily dependent on income from the oil sector. The collective actions of Britain and AIOC only made it hard for Iran to sell its oil, but did not topple the Mossadeq government. Mossadeq's downfall is blamed by many scholars on a combination of several variables: the alienation of the Iranian upper class as a result of severe measures taken by the government to collect taxes; the chaos stemming from Mossadeq's demand that the Majlis give him power to govern by decree for one year, and the refusal of the Majlis to consent to this demand; and the clashes between the Shah and Mossadeq over control of the army, which later broadened into an intense power struggle. It is a well-documented fact that Mossadeq lost this struggle with the active involvement of the U.S. Central Intelligence Agency.[17]

The restoration of monarchy in Iran also brought the restoration of the old oil concession. The consortium agreement between Iran and AIOC (which changed its name to British Petroleum Company), giving symbolic recognition to the notion of nationalization, which was still popular in Iran, reestablished the real control of AIOC over the oil affairs of that country. For the oil-producing countries in general, the outcome of the nationalization attempts in Iran was a clear signal that they had better not take that course. For the multinational oil corporations, however, the events surrounding the entire nationalization episode were forebodings of the erosion of their unquestioned control of the world oil market in the late 1950s with the entry of independent oil companies, and of the growing political and economic consciousness among the oil-producing countries which led to the creation of OPEC in 1960.

As the preceding discussion makes clear, the oil concessions were the core of conflict between the oil-producing states and the oil companies. The nature of these early concessions showed the oil states that they were not receiving their fair share of revenues. The concessions were normally negotiated between the host countries and the extracting companies before exploration had taken place. During this period the companies enjoyed, according to Theodore Moran, "near monopoly control over the resources and knowledge" the host countries needed "to develop a major tax producing, foreign exchange-producing, employment producing operation—a monopoly control that only a few fellow oligopolists could supply at a broadly similar price." Consequently, notes Moran, the initial agreement reflected the "quasi-monopolistic control" of extracting companies over the skill necessary to initiate a major operation and was heavily weighted in favor of the foreign investor. Once exploration proved to be successful, however, continues Moran, the host governments manifested resentment toward the terms of agreement with the foreign investor and labeled them exploitative. If the government in power failed to draw that conclusion, its opponents did not.[18] From the 1940s through the 1960s the oil concessions were especially unpopular. The nationalistic forces saw these arrangements as reflecting the unjust relationship that had prevailed in the colonial era. Naturally they were, in the words of Fuad Rouhani, a former secretary general of OPEC, "condemned as derogatory to the national honor" or villified because of their "close association with the regime of capitulation and colonial practices."[19]

These concessions also caused conflict because they were widely perceived within the domestic polities of oil states as "unjust." Rouhani provides three reasons for such a perception. First, these concessions were viewed as transactions between foreign extracting companies and the ruling groups, who were neither subjected to constitutional restraints nor motivated by considerations of national interest. The foreign oil companies, since they were signatories to concessions with the existing sovereign, viewed themselves as governments with governments. Acting as "quasi colonial" authorities and as "extraterritorial" regimes, these concessionaires also saw fit to interfere in the internal affairs of host countries, a role which caused considerable resentment within the nationalistic elements. The second reason, notes Rouhani, is that the host governments neither participated in the day-to-day business of the companies nor had any real control over the operations of the concessionaires. Even though the host governments were supposed to have open access to the results of the technical activities of concessionaire companies, the concessionaires were highly selective in releasing information, providing only what was self-serving and harmless. Third, Rouhani continues, the host governments authorized the concessionaires to recruit foreigners only when qualified nationals were not available, and expected the concessionaires to train nationals in order gradually to replace foreign personnel. The concessionaire companies, however, sought to postpone such recruitment as long as possible and chose to fill important positions with foreigners for an indefinite period. In this respect, observes Rouhani, "the defect did not exist so much in the provision of the concession as in the concessionaire's conduct."[20]

The Oil Companies and the Oil States

The petroleum industry comprises four distinct but related activities: exploration (which also includes production), refining, transportation, and marketing. Until the 1950s, these activities, outside the United States, were conducted either by state monopolies (operating in the Communist countries) or under government operation (e.g., Chile, Brazil, Mexico) or by the seven large multinational oil corporations.[21]

As early as in 1926, the multinational corporations were given a bitter taste of competition. In the aftermath of the Russian Revolution of 1917, the Communist government seized the property of Royal Dutch-Shell. Throughout the early 1920s, both Royal Dutch-Shell and Standard Oil Company of New York (Socony) continued to purchase Russian oil, which they sold in India and other

Far Eastern markets. After failing to negotiate a compensation from the Soviet Union for its expropriated property, Royal Dutch-Shell initiated a boycott of Soviet oil and asked Socony to do likewise. Royal Dutch-Shell was capable of recouping its loss of Russian oil by substituting oil from its concession in Romania, but Socony had no access to a similar source of supply. To oblige Royal Dutch-Shell would have weakened Socony's competitive position, a potential outcome the company was not at all willing to accept. When Socony refused to cooperate, Royal Dutch-Shell announced on September 19, 1927, that it would reduce the price of kerosene in India should any Russian oil enter the Indian ports. The price reduction was initiated when Socony proceeded to sell its Russian crude oil, thereby triggering a price war. Both companies later tried to broaden the price war to the United States and Europe. At the beginning of 1928, however, under the leadership of Socony, attempts were made to forge an international working agreement between Socony, Royal Dutch-Shell, and APOC. They finally succeeded at a grouse-shooting party at Achnacarry, Scotland, in late 1928. The Achnacarry, or "as is," agreement, underscored the willingness of Sir Henri Deterding of Royal Dutch-Shell, Sir John Cadman, chairman of APOC, and Walter C. Teagle of Socony to accept the prevailing division of the oil market and to expand production jointly. It was based upon the following six principles:

> (1) accepting and maintaining as their share of the industry the status quo of each member; (2) making existing facilities available to competitors on a favorable basis, but at not less than actual cost to the owner of the facilities; (3) adding new facilities only as actually needed to supply increased requirements of consumers; (4) maintaining for each producing area the financial advantage of its geographical location; (5) drawing supplies from the nearest producing area; and (6) preventing any surplus production in a given geographical area from upsetting the price structure in any other area.[22]

The extension of this agreement in the early 1930s to other major oil corporations, and a close observance of these principles (especially of number 6) by all participating companies, enabled them to operate as an international cartel. The cartel arrangement lasted until the early 1950s. A detailed breakdown of the magnitude of activities of the seven international corporations is provided in Table 1.

The interests of the oil-producing countries and the oligopolistic oil corporations were clearly at cross purposes and were bound to be in conflict. The *raison d'être* of the oil companies was to maximize their retained earnings by finding avenues to minimize taxes. These corporations were successful in maximizing retained earnings because they were horizontally and vertically integrated. Edith Penrose, an internationally known oil economist, contends that

Table 1. Activities of Seven International Oil Companies around 1950

	Approximate Percentage of Total Activity			
Activity	Middle East	Eastern Hemisphere	Western Hemisphere	World
Ownership of crude oil reserves	—	—	—	65
Production	99	96	45	50
Refining capacity	—	79	75[a]	57
Control of cracking capacity	—	84	53	55
Control of transportation facilities (tanker fleet)	—	—	—	50

Source: U.S. Congress, Senate, *The International Petroleum Cartel,* staff report to the Federal Trade Commission, submitted to the Subcommittee on Monopoly of the Select Committee on Small Business, 82 Cong., 2 sess. (Washington, D.C.: GPO, 1952); information compiled by the author from pp. 22-27.

Included: Five Amercian companies: Standard Oil Co. of New Jersey; Standard Oil Co. of California; Socony Vacuum Oil Co.; Gulf Oil Corp.; Texas Co. Two British-Dutch companies: Anglo-Iranian Oil Co.; Royal Dutch-Shell group.

[a]Exclusive of the United States.

a high degree of integration enabled the oil companies to be arbitrary in allocating costs to different operations and in setting the prices for the transfer of goods and services between subsidiaries. She believes that the international oil firms juggled their overhead costs and adjusted transfer prices among their foreign branches and affiliates in such a way as to reduce their total tax outlays. The behavior of the multinational oil corporations contrasts sharply with the behavior of the oil states. The latter were concerned with maximizing their taxable earnings, but their knowledge of tax laws was too primitive to match the sophisticated tax-evasion techniques employed by the oil corporations, especially before the formation of OPEC.[23] In addition, the oil corporations used highly questionable tactics, according to Franklin Tugwell, such as "playing producing countries off against each other where practicable."[24]

The *modus operandi* of the multinational oil corporations bred suspicion within the host countries. The oil companies, like all large firms, had the upper hand over their competitors, as well as over the host and parent governments, largely because of the secrecy surrounding their business decisions. In addition, these corporations owed first allegiance either to themselves or to their stockholders, so their decisions were aimed at maximizing their own or their stockholders' gains.[25] Such decisions were bound to come into conflict with some vital national interests of the host countries, thus increasing the need for operational secrecy on the part of the multinational corporations. The host countries, meanwhile, remained deeply distrustful of the corporations and their motives. There was little assurance that the outcome of the corporations' economic activities would be consistently beneficial as long as the corporations were not held under tight control. Until the early 1970s the oil states had no such control over the behavior of the oil companies. Moreover, the secrecy system under which these companies operated left the policymakers of the oil states with enormous uncertainty and with an uneasy but more or less certain feeling that, in the words of Tugwell, "given the need and opportunity, their country's welfare would be sacrificed by externally-based decision-makers."[26]

The relationship between the oil-producing states and the multinational companies was destined to produce chaotic results as soon as the balance of power shifted away from the oil companies. The international oil market reflected this shift when it converted from a buyers' to a sellers' market in the 1970s.

Plan of the Book

The next six chapters examine the dynamics of the uncertainties of oil power as they influenced the behavior of, and in turn were influenced by, the oil-producing nations, the oil companies, and the industrial consuming countries from 1960 through 1985.

Chapter 2 describes the uncertainties of the buyers' market that persuaded the oil-producing countries that only by putting forth an organizational front could they cope with the mighty multinational oil corporations. OPEC was meant to be such an organization. Chapter 3 deals with the transformation of the international oil market from a buyers' to a sellers' market, and uncertainties attendant on that change. Chapter 4 examines the chaotic results of the acute uncertainties of the sellers' market, which emerged in the form of quantum leaps in prices and insecurity of supply. The behavior of oil-consuming and oil-producing nations in this period illustrated the maximization of their respective national interests. Chapter 5 analyzes the Arab oil embargo, which was imposed by OAPEC in October 1973 and was lifted in March 1974. The oil embargo underscores the theme of this study: that the pricing behavior of the oil states was essentially economic in nature. Chapter 6 evaluates the oil market from 1976 through the first quarter of 1985, a period during which OPEC as an organization experienced a number of ups and downs. That chapter also discusses a variety of challenges emanating from the oil market of the 1980s, which is significantly different from the oil market of the 1970s; from tensions stemming from the war between Iran and Iraq; and from the growing political rivalry between the Khomeini regime and the Persian Gulf oil states. Chapter 7 puts the past twenty-five years of OPEC in perspective and examines the challenges the organization faces in the coming years. The Appendix contains statistical data.

2. The Uncertainties of the Buyers' Market

The objective realities of the 1950s persuaded the oil-exporting countries that only by establishing a collective and organized front could they keep the mighty oil companies from introducing unilateral decreases in the price of their commodity. This realization fueled their endeavors to create OPEC in 1960. The inception of OPEC was no guarantee that the organization would be a powerful one, however. The uncertainties of the buyers' market that prevailed between 1960 and 1967 prevented OPEC members from resolving three important policy issues—stabilization of oil prices, institutionalization of prorationing, and expensing of royalties. The year 1967 has been selected as a cutoff point. Because this was the year of the third Arab-Israeli war, which ultimately resulted in a qualitative change in the bargaining position of OPEC, it requires separate treatment.

The Genesis of OPEC

OPEC was not formally established as an economic alliance until September 1960, but the original contacts between the oil-exporting countries of the Middle East and Venezuela, and other activities leading to its creation, date as far back as the 1940s. During that decade the discontent generated by the regime of oil concessions brought about the conviction in the oil-producing countries that the status and obligations of foreign operators must be subjected to a new order.[1] Hence, the Arab League, founded in 1945, entertained the idea of creating a petroleum association of Arab countries which would also include the large non-Arab exporters of petroleum, notably Iran and Venezuela.[2]

As an initial step toward coordinating their oil policies, the governments of Iran and Venezuela established diplomatic contacts in 1947, although not until 1949 were these contacts formalized. An official Venezuelan mission visited

Iran, Iraq, Kuwait, and Saudi Arabia in 1949 to exchange views and explore new avenues for closer and more regular communication among them in the future.[3] The real reason for this spurt of activity on the part of Venezuela, however, was apprehension over the growing competition from very cheap oil from the Middle East. The Venezuelan government hoped to induce the Middle Eastern governments to raise fiscal charges on concessionaire companies. The Middle Eastern oil countries, through such contacts, acquired a considerable amount of technical information. In fact, tax-related information provided by Venezuela to Saudi Arabia, Iran, and other governments of the Middle East during 1947 and 1950 was partly responsible for increasing the level of payments by oil companies to host governments.[4] These formal contacts also led to a realization that such cooperation might be financially very rewarding. Such feelings paved the way for sustained and exhaustive cooperative endeavors, which eventually led to the establishment of OPEC.

The next significant formalized contact was the Iraqi-Saudi agreement of 1953. This agreement was signed at a time when the international oil companies, because of the abortive nationalization of the Iranian oil industry in 1951, had demonstrated an impressive show of cooperation by successfully boycotting Iranian oil. The Iraqi-Saudi arrangement was the first formal agreement of cooperation between oil-exporting countries aimed at exchanging oil information and holding periodic consultations about oil policies in order to improve their bargaining position vis-à-vis the companies. A noteworthy feature of this agreement was the "best-term clause," which enabled the host governments to call on their concessionaires to discuss possible revisions of agreements if neighboring countries could obtain better terms.[5]

While the oil-exporting countries were becoming increasingly conscious of the necessity for cooperation and exchange of information, other forces were also emerging in the international oil market, the cumulative result of which was erosion of the oligopolistic power of the international oil majors. This erosion of power meant that the oil-exporting countries were able to obtain more maneuverability and discretion in choosing to whom they would sell their commodity. In addition, they were to be in a slightly more advantageous position to bargain for better prices. Although these forces may not be directly responsible for the establishment of OPEC, the fact that they helped loosen the iron grip of these companies on the international oil market made it easier for the oil states to continue a series of activities which culminated in the creation of OPEC.

The period following World War II found a number of sizable American firms crudeshort and anxious to enter the foreign market. Domestically, the price of U.S. crude was maintained well above the world price by restricting domestic production and by enforcing prorationing.[6] But U.S. tax laws encouraged overseas investment by allowing a substantial portion of payments to foreign governments to be used to offset American taxes. Thus, independents were afforded enormous economic incentives to enter the major oil-producing areas.

The entry of independents into the international oil market had several implications. First, the willingness of new entrants to pay a much higher price raised the consciousness of the host governments to the fact that they had been shortchanged by the international majors, and the resulting competitive bidding increased the price of concessions. Second, independents were responsible for the expansion of world supply, which led to erosion of prices. When the international majors attempted to curtail production to preserve the price structure, they met with resistance by host governments.[7] Another noteworthy implication of the entry of independents was their ability to develop downstream outlets for foreign crude oil, thereby loosening yet another part of the majors' control of the market. The international political environment also proved to be auspicious for the success of independents. With the decline of colonial powers (e.g., U.K. and France) which, as mother countries, had supported the majors, the majors no longer remained on firm ground. The United States began not only to insist that the majors open up their transactions but to require that independents be allowed in some operations.[8] The modernization efforts initiated by the oil-producing countries also had an impact on the independents, for these endeavors required additional revenues, and the independents proved to be convenient and willing sources for these rapidly growing capital needs.

The entry of the Soviet Union into the international oil market in 1953, like the entry of the independents, contributed to loosening the hold of the international oil companies. Although the fact was not immediately apparent, the Soviet Union helped to create a favorable environment for the oil-exporting countries. That country's resolve to expand exports enabled it to sell at cut-rate prices or through barter. To head off competition from the USSR, as well as that from small independent producers, the major oil companies started offering cash discounts and other inducements. The discounts varied between 10 and 60¢/b on Middle East crude and were occasionally raised to as much as 30 percent of posted prices for spot cargoes.[9]

Another significant force for change was the emergence of state-owned oil companies, whose main purpose was to increase indigenous participation and control of the industry. These companies began to enter joint ventures with the independent oil companies, creating two unintended results. These endeavors increased competition. And, because more oil was produced as a result of these joint ventures than could be readily absorbed by the international market, they further contributed to reductions in price.[10]

Finally, the international majors' behavior during the 1950s was itself responsible for the sharp decline in oil prices toward the end of that decade:

> During most of the 1950s, and especially after Suez, the major companies seem to have used a variety of techniques to grant concealed discounts on

> the effective c.i.f.[11] price of crude oil in a number of markets where the competition to sell products was particularly keen. The techniques included freight allowances on delivered prices, the re-charter of tankers at favorable rates to f.o.b. buyers, extra quality concessions, "spiking" to uplift the quality of crudes without extra charge, the long-term financing of crude purchases at favourable rates, and acceptance of partial payment in soft currencies by the U.S. companies.[12]

The net result of all these factors was not only erosion of the oil majors' oligopolistic control of the international market but also a sharp decline in the price of oil.

The Iraqi revolution of 1958, which abruptly ended twenty-six years of the Hashimite monarchy, was also the beginning of a number of significant oil-related activities. This revolution brought to power the nationalistic government of Brigadier Karim Qassem. His regime felt that, through united Arab efforts, Iraq might extract a better deal from the Iraq Petroleum Company (IPC), 23 percent of which was owned by British Petroleum, the Royal Dutch-Shell Group, and CFP; 11⅞ percent, by Standard Oil of New Jersey and Mobil Oil; and 5 percent, by the Gulbankian interests. In Saudi Arabia, King Faisal consolidated his political hold, and, as a result of a rapprochement with President Gamal Abdel Nasser of Egypt, Saudi Arabia could concentrate on oil matters. Abdullah Tariki, the Saudi Arabian director general of Petroleum and Mineral Affairs, initiated demands for greater Arab participation in the petroleum industry and for greater concessions from the foreign companies.[13] These events clearly manifested heightened consciousness even on the part of those oil-producing countries that ranked low in political, social, and economic development.

By January 1959 the new Iraqi government was calling for some united Arab action. The government of Iraq pointed out that even though the international companies were supposed to be competing with one another, in reality they were pursuing a uniform policy toward the Arab governments while the Arab governments were following separate and different policies toward the companies, policies that worked against their own interest. To remedy this situation, Iraq suggested that the Arab governments arrive at a uniform policy for the exploitation of their resources.[14]

One month later, the oil companies announced reductions in posted prices of 5-25¢/b for Venezuelan crude and of 18¢/b for Middle Eastern crude.[15] In April, the First Arab Oil Congress met in Cairo, with Venezuela and Iran as observers. Iraq was conspicuously absent because of an ongoing political feud between Qassem and Nasser. The Congress offered a handy forum for expressing the dissatisfaction and apprehension of the oil-exporting countries over these reductions, but, by terms of their agreements, concessionaires had no obligation to consult Middle Eastern governments on changes in posted prices.

These informal discussions between the Arab and Venezuelan representatives resulted in a consensus on the utility of collective action in settling common grievances with the oil companies. The conferees also agreed "on the desirability of their governments' jointly establishing a consultative commission to meet annually and discuss and seek a solution to the problems of mutual interests."[16] The most significant steps taken by the Congress were passage of a resolution calling for alterations in the profit-making formula and the demand that no change in the oil price structure or in the prices themselves take place without prior consultation with the governments of the oil-producing countries.[17]

In August 1960, when the oil companies announced a new series of price cuts, the exporting states reacted decisively. Representatives of Iran, Iraq, Kuwait, Venezuela, and Saudi Arabia met from September 10 through 14 in Baghdad and laid the foundation of OPEC.

The creation of OPEC was thus the outcome of a variety of forces and factors. The numerous contacts among oil-exporting countries made them increasingly aware of their mutual interests and heightened their sense of the need for cooperation. The growing uncertainties among the oil-exporting countries during the 1950s occasioned by the downward slide in the price of oil, in conjunction with what they perceived as unwarranted and unilateral decisions by the oil companies to introduce price cuts, led them to believe that the only way to deal with the awesome organizational power of the multinational oil corporations was to establish an organization of their own. Such an organization, they hoped, might serve to stabilize the prices of oil.

Thus, OPEC came into existence as an economic alliance, with Iran, Iraq, Kuwait, Saudi Arabia, and Venezuela as its founding members. Qatar joined in 1961; Indonesia and Libya in 1962; Abu Dhabi in 1967 (Abu Dhabi's place has since been taken by the United Arab Emirates [UAE], of which Abu Dhabi is the dominant member); Algeria in 1969; Nigeria in 1971; and Ecuador in 1973. Gabon was admitted as an associate member in 1973.

The Institutional and Behavioral Setting

A broad overview of OPEC's organizational structure and the functions of its various bodies will provide a better understanding of its decision-making process. As stated in Resolution I.2, paragraph 7, of OPEC's statute, membership is open to any country "with substantial net export of crude petroleum," but admittance requires the unanimous consent of the founding members. The membership is divided into three categories: founding members, full members (including the founding members), and associate members. According to Article 7 of OPEC's statute, in order to become a full member of the organization, a petroleum-exporting country must have "fundamentally similar interests to

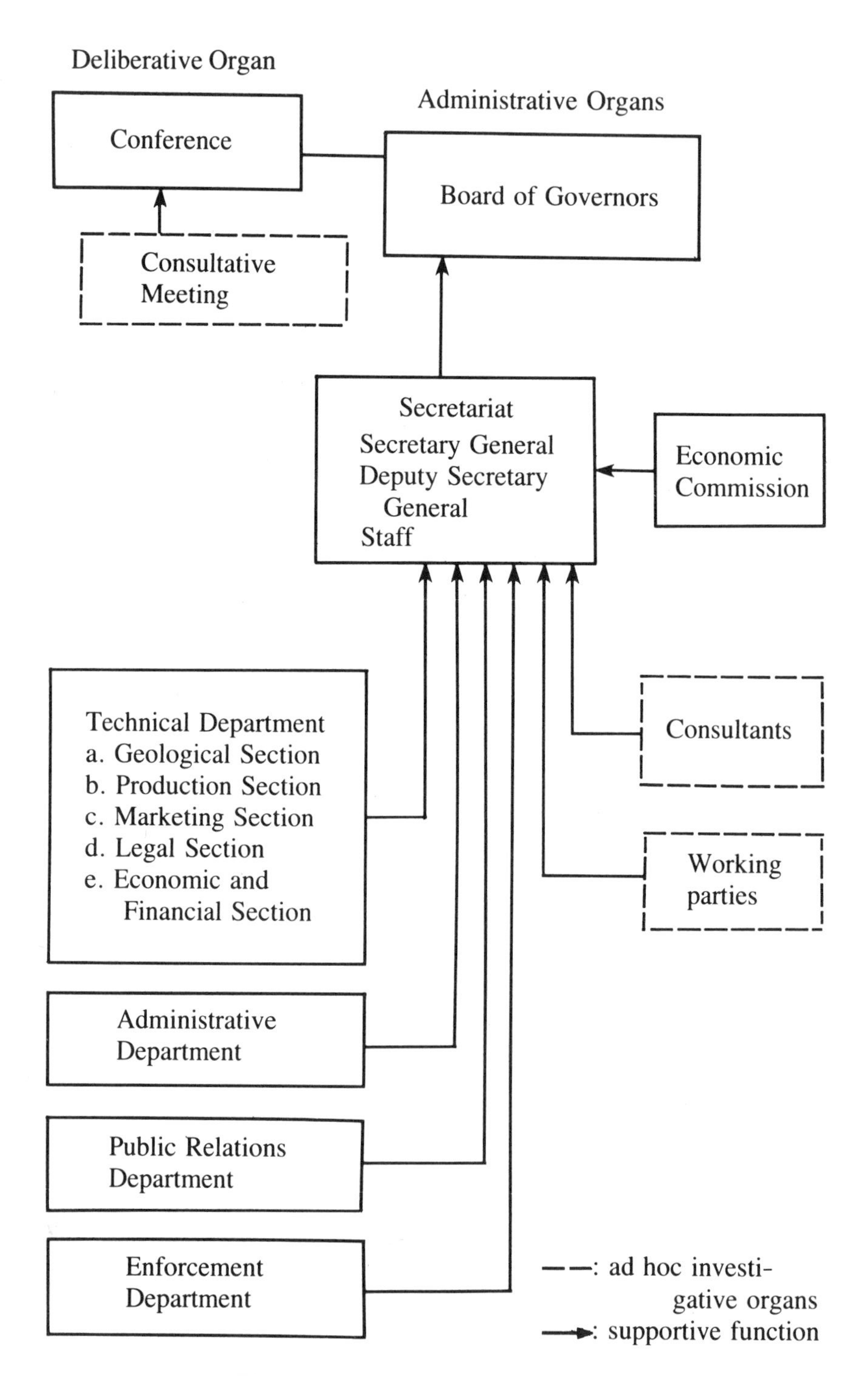

Figure 1. The Organization of OPEC

those of member countries." OPEC is composed of four permanent organs, the Conference, the Board of Governors, the Secretariat, and the Economic Commission. In addition, the organization frequently utilizes such ad hoc investigative arrangements as consultative meetings, consultants, and working parties (see Figure 1).

The Conference, which is made up of representatives of member countries, is OPEC's "supreme authority." It holds Ordinary Meetings twice a year. At the beginning of each Ordinary Meeting, the Conference elects a president. An Extraordinary Meeting of the Conference may be called at the request of the Board of Governors or the chairman of the Board. All the decisions taken by the Conference, save those involving procedural matters, require the unanimous consent of the full members.

All matters of detailed examination are referred by the Conference to a Consultative Meeting, which consists of the representatives of the member countries and is an ad hoc investigative body. When the Conference is not in session, the president may also refer matters of urgent consideration to the Consultative Meeting. The Consultative Meeting, after studying the referred matters, submits its findings and recommendations to the Conference for approval.

The following functions are performed by the Conference: (1) formulating general policies for OPEC, and determining appropriate means of implementing them; (2) deciding on new membership and confirming the appointment of members to the Board of Governors; (3) directing the Board of Governors to submit reports, making recommendations on matters relevant to OPEC, and taking decisions on these reports and recommendations; (4) considering and approving the budget, the statement of accounts, and the auditor's report submitted by the Board of Governors; (5) calling a Consultative Meeting when necessary; and, (6) appointing the secretary general, the deputy secretary general, the chairman of the Board, and the auditor.

The Board of Governors is charged with directing the management of the affairs of OPEC and implementing the decisions of the Conference. Initially, each founding member was entitled to nominate one governor, while the new members collectively were to appoint one governor. This discrepancy was later corrected, however, and all members are represented individually at the Board meetings. The Conference appoints one chairman of the Board from among the governors for a period of one year based upon the principle of rotation. Originally the chairman of the Board also served as the secretary general of OPEC, but in May 1965 this dual role was abolished and the chairman of the Board is now appointed separately for each year.

The Board of Governors holds at least one meeting every three months at the OPEC headquarters in Vienna. A quorum of two-thirds of the governors is required to hold a meeting, and the decisions are made on the basis of a simple majority of attending governors. Whenever a majority of governors determine that the continued membership of any governor is detrimental to the interests of the organization, he is suspended from the Board and his replacement is approved by the Conference. The Board of Governors is also responsible for convening Extraordinary Meetings of the Conference, preparing the agenda for the Conference, and nominating a deputy secretary general for appointment by the Conference.

In order to organize and administer the work of the organization, Resolution I.2, paragraph 6, called for the establishment of a Secretariat. This body is headed by a secretary general, who is appointed for a period of one year. As the chief officer of the Secretariat, he conducts the affairs of OPEC under the direction of the Board of Governors. The deputy secretary general is selected by the Board of Governors from a list of qualified nominees from member countries. He serves for a period of three years, and his term may be extended for an additional year pending the suggestion of the Board of Governors and approval of the Conference. He assists the secretary general in coordinating the technical and administrative activities of the Secretariat, and acts as secretary general in his absence. The Secretariat staff is appointed by the secretary general and with the approval of the Board of Governors. Article 31 designates this staff as "exclusively international." The Secretariat staff is chiefly concerned with promoting the interests of OPEC. The functions of the Secretariat include: (1) collecting information, studying and reviewing all matters of common interest relating to the petroleum industry, and reporting to the Board of Governors; (2) making specific studies and preparing proposals whenever requested by the Board of Governors; (3) proposing specific measures for possible adoption by the Conference; and (4) implementing the decisions taken by the Conference and providing technical guidance and assistance to member countries in the implementation of these decisions.

The secretary general is supported by a technical department, an administrative department, a public relations department, and an enforcement department. The technical department is charged with reviewing the global petroleum situation and its implications for OPEC, and submitting reports to the Conference through the Board of Governors. The technical department is made up of a geological section, which conducts geological studies for OPEC; a production section, which collects data on reserve and production potential for major producing countries and other potential producers; a marketing section, which studies the world energy markets and forecasts consumption trends; a legal section, which is charged with studying the legal aspects of the oil business and advising member states on these matters; and an economic and financial section, which assists OPEC members on a variety of financial matters having to do with the oil industry. The Public Relations and Enforcement departments are two additional organs which support the secretary general by disseminating information and by implementing the decisions of the Conference, respectively.

The Economic Commission was established by the Seventh Conference of OPEC as an advisory organ. It is comprised of a board, a staff, and national representatives. These representatives, who are appointed by the member countries, serve as permanent liaison officers with the Commission. The Economic Commission functions within the general framework of the Secretariat. It collects relevant economic data from a variety of sources in order to prepare recommendations for individual member states. It should be noted, however, that the Commission is dependent on OPEC members for the accumulation of statistical information on their respective oil affairs. This body also assists the secretary general in preparing reports on the condition of the world petroleum market, in formulating recommendations to forestall downward slides in the prices of crude oil and crude products, and in determining whether the issues involved need the urgent consideration of an Extraordinary Meeting.

Resolution VI.46 gives the secretary general the discretionary authority, whenever necessary, to commission consultants and working parties to conduct studies that cannot be undertaken by the Secretariat, or to carry out studies on specific subjects of interest to the member countries. These arrangements are ad hoc in nature.

OPEC's most visible decisions related to the price and production rates of crude oil are taken by the Conference. Since these decisions affect the pace of economic growth in the non-Communist world, they routinely receive extensive coverage by the international press. The other organs of OPEC carry out their mundane chores free from such publicity.

The Initial Challenge

The issues faced by the young OPEC were quite complicated and the challenges were immense. Three outstanding issues were the stabilization of oil prices, institutionalization of prorationing, and expensing of royalties. A brief discussion of each of these is in order.

Stabilization of Oil Prices The issue of stabilization of oil prices, which was a very significant objective of OPEC, suffers from a lack of definitional precision. What precisely did the term mean? Stabilization may be used to mean that OPEC members wanted to forestall any further downward slide in the prices of crude oil akin to the two price cuts unilaterally introduced by the international oil corporations prior to the creation of OPEC. Thus stated, the stabilization of prices was an objective clearly attained by OPEC during the period under discussion, since the price of oil remained frozen at the post-August 1960 level. On the other hand, if price stabilization is used to mean efforts by OPEC members to restore oil prices to the pre-August 1960 level, then OPEC's record is problematic and requires a closer look.

OPEC members, in the maiden resolution of their First Conference, explicitly stated that they would "devise ways and means of ensuring the stabilization of prices in the international oil markets" in an attempt to eliminate "harmful and unnecessary fluctuations." They also emphasized that future price modifications should be introduced only after prior consultation with the oil-producing states and that the oil states should employ all means available to them to restore oil prices to the pre-August level. The same resolution also declared that if a member country was penalized by the oil companies "as a result of the application of any unanimous decision of this conference," other members should refrain from accepting "any offer of a beneficial treatment . . . which may be made to it by any such company or companies."[18]

An additional strategy was "the adoption of a classical measure to stabilize prices—namely the establishment of control over supply." The problem with this measure, however, was what criterion was to be employed to control production. For example, if the population of an oil-producing state was the criterion, then Iran and Iraq would get preference over Saudi Arabia and Kuwait; if the share of oil revenues in each national budget was the criterion, then Saudi Arabia and Kuwait would be favored over the former two countries. "If prorating of production should be adopted in order to maintain high prices for crude, exploration and extension of exploration or reserves would practically come to a standstill in the pro-rated countries and, conversely, would expand in the non-pro-rated countries, even though the lands of OPEC members contained a very high percentage of the world's reserves."[19]

As a negotiating technique, the member countries chose individual country-company negotiations over collective bargaining. This was a major tactical error, "since it undermined at an early stage the effectiveness of any collective bargaining that the producing countries might have hoped for." In addition, considering the fact that the world market was characterized at the time by a surplus of crude oil, it was scarcely to be expected that the oil companies collectively could be persuaded to surrender their administrative discretion on pricing decisions. Another mistake that may have undermined OPEC's objective was its announcement in November 1962 that member countries were unanimous in emphasizing the need to conclude the negotiations by March 30, 1963. Should they fail to achieve this goal, the statement added, they "would consult with a view to taking appropriate steps for the implementation of these resolutions." It is significant to note that when the negotiations failed, OPEC also failed to live up to its own commitment to take whatever appropriate steps it may have had in mind. This failure caused serious damage to its credibility.[20]

The inability of OPEC to achieve its central objective—the stabilization of prices—had a great deal to do with a variety of actions taken by its members outside the framework of that organization, which, instead of curbing the surplus of crude oil in the international market, resulted in further quantitative increases in the supply situation.

In order to break the oligopolistic control of the international oil companies, the oil-exporting states adopted a general policy of reclaiming unexplored acreage from concessionaries and selling it to the new independents for a substantial price. This policy was intended to confer two advantages (aside from increased income) on the oil states. First, the exporting countries expected that the threat of the loss of unexplored territories to the independents would force the major concessionaires to expand the zones of their activity and thereby to increase production. Second, the oil states hoped that the introduction of independent oil companies in the area would mean they would have more customers from which to choose. Additional customers also would mean added economic leverage for the exporting countries. An unintended effect of this policy was expanded production, and this proved to be detrimental to price stabilization.

Another undermining factor was the role played by the national oil companies owned by the oil-exporting countries. These companies were originally established to increase gradually the indigenous control over the management and production of oil at home and to augment the government's role in the downstream operation of the oil industry. These companies, through negotiations with their major concessionaires, acquired large quantities of crude oil, but, since they were not vertically integrated, they had to sell this oil to the companies that owned downstream facilities (transportation, refining, distribution, and marketing facilities). Another reason for this type of sale was that the national oil companies, because they were seeking a foothold in the international markets and wanted to establish themselves as reliable sources of supply, were willing to offer potential buyers attractive terms, mostly in the form of price discounts.[21] The net result of the activities of the national oil companies was an increase in crude supplies, which had already surpassed the demand. A concomitant and inevitable result of this was further weakening of the prices of crude oil.[22]

The Fourth Conference of OPEC, which was held at Geneva in April 1962, reflected a more determined tone and a more restricted aim. Three important resolutions were passed by the conferees. The first one (IV.32) called for formal negotiations with the oil companies for the restoration of pre-August 1960 prices. The other two called for the expensing of royalties (IV.33)—asking the concessionaires to treat royalties as a cost instead of setting them as a credit against income tax—and elimination of marketing expense[23] by the concessionaires (IV.34). Of these resolutions, OPEC eventually attained its objectives on the latter two, but the issue of rescinding the August 1960 price reduction was the most difficult to settle. It took OPEC ten years (from 1960, the year of its establishment, to 1970, when the international oil market was in the process of transformation from a buyers' to a sellers' market) to accomplish this objective.

Prorationing From the time of the establishment of OPEC, two representatives who shared remarkable farsightedness, a Saudi and a Venezuelan, Shiekh Abdullah Tariki and Dr. Perez Alfonzo, were convinced that the most effective *modus operandi* to obtain restoration of pre-August 1960 prices was to control production, a technique otherwise known as prorationing.

The effectiveness of this measure as a means of restoring or stabilizing prices had its proponents and its opponents. Its proponents argued that any attempt to protect the price level not directly linked to the factors of supply and demand would be doomed to failure.[24] The opponents, on the other hand, felt that the institutionalization of prorationing by the oil-exporting countries would create an enormous amount of ill will among consuming nations against the oil states or might even trigger a serious response, such as a boycott by the oil companies. With memories of the multinational oil companies' successful boycott of Iranian oil in the early 1950s fresh in mind, the opponents of prorationing were not willing to take any unnecessary risks associated with the technique. Even if OPEC members agreed to incorporate this measure to stabilize or increase the price of crude oil, they would have to resolve still other problems associated with it. If all the producers limited their production and attained price stabilization or increases, it would become particularly tempting to one of the participating countries to withdraw from the association in order to produce at liberty and to sell at the most advantageous price. Another problem was the criteria for allocating the production quotas to various members. Possible criteria included "the volume of proven reserves, the rate of current production and the prospects of its annual growth, the size of investments, the costs of operation, the population, the proportion of oil exports to total exports, and the proportion of oil revenues to total revenues."[25]

The importance of prorationing was debated at the First Arab Congress in 1959, before the creation of OPEC. For two consecutive years after the establishment of this organization, its membership acknowledged the importance of prorationing, once in 1960 in its first resolution and again in 1961 by emphasizing the need for a just price formula supported by international proration, should that prove essential.[26] Venezuela's pursuance of this matter continued even at the Third Arab Congress in 1961.

Venezuela was the only country that clearly understood the importance of prorationing and remained the most enthusiastic advocate of using this measure to restore prices to the pre-August 1960 level. During the late 1950s the Venezuelan rate of production suffered setbacks because of increased production in the Middle East. Being familiar with the American experiment in price stabilization through oil proration and confronted by declining production rates and loss of revenues through price cuts, the oil minister of Venezuela, Dr. Perez Alfonzo, began a campaign for the institutionalization of prorationing by OPEC. Realizing, however, that outright acceptance of prorationing by the Middle Eastern oil states would be almost impossible to obtain, he focused on the "lesser objective" of convincing them that they should at least follow the Venezuelan example and establish coordinating commissions to monitor company activities and assert the state's role in marketing.

The oil-exporting countries of the Middle East, in contrast, were not very receptive to prorationing. Perhaps one of the primary reasons for this was that these governments were poorly prepared and poorly informed about petroleum policy and were not capable of effectively utilizing this strategy until they stopped depending entirely on the oil corporations for information.[27] In addition, the adoption of prorationing meant that the oil states had to alter their rate of production in response to upsurges and downsurges of the oil market. It also meant lowering production when necessary. Such a measure would have come into conflict with the domestic policies of all oil states save Venezuela. The Middle Eastern oil states also remained unpersuaded that prorationing would either stabilize prices or push them upward. "As OPEC's enthusiasm for devising a prorationing program waned, Venezuela's waxed." On one occasion Venezuela reportedly threatened to resign from OPEC if it "did not tackle the problem of output control. Tardily it did so."[28] The result was Resolution VII.50, passed in November 1964 at the Seventh Conference, which expressed OPEC's concern over the continuous decline of crude oil prices. It was felt that in order to determine the basis of the organization's policy, a specialized and continuing study of the market was more than ever necessary.[29] Thus, OPEC decided to set up an Economic Commission as a permanent special organization whose purpose was to examine price fluctuations in the international oil market and to formulate recommendations to counteract continuing erosion of crude and product prices, including, if necessary, a production control program as contemplated in the initial Baghdad resolutions. The commission was not formally established until the Eighth Conference, held in Geneva in 1965, but in the meantime it had worked out a proposed schedule of allowable output increases for the OPEC region as a whole and devised a plan for dividing the increases among its members. These measures, when presented for discussion at the Geneva conference, apparently precipitated such sharp dissension and bitterness, however, that OPEC took no formal action on the Commission's recommendations.[30]

At the Ninth Conference meeting in Tripoli in July 1965, OPEC adopted as a "transitory measure" its first production program.[31] Though no actual proposals for curtailment of production were made, the conference, on an experimental basis, formulated limited percentage increases for 1966 as follows:

Venezuela, 4 percent; Indonesia, 4 percent; Kuwait, 5 percent; Qatar, 6 percent; Saudi Arabia, 9 percent; Iraq, 9 percent; Iran, 16 percent; and Libya, 33 percent.[32] The member countries, however, ignored these limited percentage increases. Libya, whose oil production was on the increase, declared that it would observe no production limit that had been imposed, and never would accept one. Libya announced, moreover, that it intended to develop production until it reached maturity. Saudi Arabia expressed skepticism over the effectiveness and workability of a proration program that was limited to OPEC areas, and accepted it for only six months on a strictly experimental basis.[33] Iran, which was preoccupied with increasing its oil revenues and regaining its pre-nationalization level of production, criticized prorationing as a program that would merely expedite the expansion of alternate sources of supply and would fail to stabilize prices.[34] By June 1967 OPEC had lost all hope for the immediate implementation of prorationing as a measure to stabilize or increase prices, although it did not cease to pay lip service to the concept of prorationing.[35]

Expensing of Royalties At the time of OPEC's formation, the oil companies paid royalties to the host governments for the right to explore, to pump oil, or to put the oil on the market. The arrangements for royalty payments to a number of oil states at that time were as follows:

> Iran, Kuwait, Libya and Qatar were entitled to 12.5 percent of the value of the oil produced, on the basis of posted prices; Saudi Arabia received under this title 22¢ per barrel, which was in effect equivalent to 12.5 percent of the value of the oil produced. Indonesia had excluded royalties from the scope of its petroleum legislation, the concessionaires being required to pay to the state, in satisfaction of all financial and fiscal obligations, 60 percent of their profits. The only OPEC member country that benefitted from a rate of royalty higher than 12.5 percent was Venezuela. The rate applied in this country was 16⅔ percent.[36]

The organization took its first step to correct this inequity during the Fourth Conference in 1962, when it adopted a resolution (IV.33) demanding that royalty payments be treated as a cost, not as a credit against the companies' tax obligations to the host countries. This action marked the beginning of three years of intricate negotiations and tough bargaining.

As expected, the initial reaction of the oil companies to this demand was outright rejection. Since the establishment of OPEC, the international oil companies had maintained an attitude marked by suspicion and skepticism toward the young organization. They perceived OPEC as inevitably and intrinsically hostile to their interests and had little doubt concerning its survivability as an economic alliance.

The negotiations on royalties turned out to be long. Throughout the negotiations the oil companies offered a series of responses that were aimed at gradually accepting the position of OPEC. At first the companies persistently rejected the principle of collective bargaining, a technique they had used so skillfully in negotiating with the petroleum-exporting countries on numerous occasions. Then the companies categorically rejected the position of the oil states that the expensing of royalties was problematic and needed negotiation. Finally, the companies conceded that the issue of royalties needed rectification and offered specific terms. OPEC characterized these as meeting its minimum demands, and referred them for action to its member countries.[37]

OPEC asked Iran and Saudi Arabia to conduct direct negotiations with the concessionaires on behalf of all the members, since these two countries maintained smooth relations with the companies. Later OPEC commissioned Fuad Rouhani, a former secretary general of OPEC, to conduct the negotiations as its representative. He was later replaced by a three-man negotiating team, and eventually the royalty issue was referred back to the individual countries.[38] A breakthrough was reached when the consortium of oil companies in Iran and some of the other companies made an initial compromise offer. The companies were willing to recognize the principle of expensing of royalties if the oil states would allow, for tax purposes, a discount below the posted prices. OPEC rejected this proposal as unsatisfactory.[39] Although no satisfactory or acceptable solution emerged from these negotiations, the oil states won a symbolic victory because the companies legitimized the principle of royalty expensing by making counterproposals. Finally, in the middle of 1964, an agreement was reached. Under it, the expensing of royalties was agreed upon, but the governments allowed, for tax purposes only, discounts of 8.5 percent for 1964, 7.5 percent for 1965, and 6.5 percent for 1966.[40]

OPEC as an organization never gave its collective approval to this agreement. At the Seventh Conference, held in Jakarta, Saudi Arabia, Iran, Kuwait, Qatar, and Libya argued that the revised offer of the companies was the maximum that could be obtained without coercion, and that they were not in favor of applying coercion.[41] Iraq opposed the compromise and was supported by Venezuela and Indonesia, two member governments that had not been directly involved in these negotiations.[42]

The oil states continued to search for a formula to eliminate the discount allowance. The 1967 Arab-Israeli war, which resulted in closure of the Suez Canal, also introduced some significant changes in the production pattern of Middle Eastern oil and in oil company-government relations. In January 1968 the oil companies made an offer to phase out the remaining 6.5 percent allowance over a period of seven years.[43] In the absence of a unanimous agreement over the companies' new offer and as a face-saving device, OPEC then passed another resolution stating that "the acceptance or rejection of the royalty settlement should be left to the Member Countries concerned."[44]

The pendulum of oil power was clearly swinging in favor of the multinational oil corporations. Throughout 1960-1967 the companies were riding the uncertain tide of the buyers' market, and OPEC remained a weak and ineffective institution. The challenges faced by the fledgling organization—price stabilization, prorationing, and expensing of royalties—were enormously complicated and were frequently in conflict with the economic interests of a number of member states. Almost all members save Venezuela lacked the kind of technical expertise required for dealing with the international majors, who not only excelled in every aspect of the oil business but also, despite a flurry of activity by the independents, retained firm control over the international oil market.

The uncertainties of the buyers' market played a key role in keeping the oil-producing countries from developing consensual plans of action. For example, price stabilization and prorationing as policy objectives were mutually complementary in nature. But the implementation of these policies presented serious problems. The chief concern of the oil-producing countries of the Middle East during this peroid was to advance their oil industries and increase oil revenues by raising their rate of production. They were looking for additional customers to buy their oil and for new partners with whom to initiate ambitious joint ventures. Under such circumstances it was quite unreasonable to expect them to be enthusiastic about adopting prorationing, which would have required them to lower production in response to fluctuations in demand and supply in the international oil market. Prorationing suited only the policy objectives of Venezuela, whose oil industry was the most developed among the OPEC members and whose oil had lost its competitive advantage because of the low cost and increased rate of production of Middle Eastern oil.

The effective implementation of prorationing also necessitated the existence of an autonomous agency, "able and equipped with sophisticated computers, to match efficiently the complexity of crude oil supply with a still more complex pattern of demand for petroleum products."[45] Such an agency also needed the authority to enforce its decisions on the recalcitrants with sanctions. This type of arrangement was well nigh impossible to bring about among OPEC members. OPEC was an organization of sovereign and independent states which agreed only to hold regular consultation with a view to coordinating and unifying their policies, and to take individual or collective actions aimed at best promoting their mutual interests.[46] The member states retained all rights to dissent from OPEC and to implement policies that they perceived to be in their own interests. In this instance, the members not only rejected the prorationing formulae suggested by OPEC but also pressed their concessionaire companies "to increase production to the utmost and to raise their share of the world oil trade." In addition, almost all member nations continued to open new areas for exploration and development. All this was "incompatible with OPEC's professed objective of containing production."[47]

During the course of negotiations on the royalty expensing issue, there was no agreement among OPEC members concerning the nature and scope of their demands. It is possible that this uncertain and shaky posture had something to do with lack of experience, since this was their first encounter as an organization with the multinational companies. At first, the companies refused to recognize OPEC and adopted an attitude of indifference toward its purpose. Even though in 1962 they expressed a willingness to acknowledge OPEC's existence as an intergovernmental agency, they remained "extremely anxious to avoid negotiating directly with OPEC" because they did not want "to magnify the importance of what they liked to think of as simply a forum for discussion and an advisory information bureau for oil governments."[48]

As the advantaged actors, the oil companies were in no mood to give up their decisive advantages over the oil states. The December 1964 royalty-expensing offer made by the companies exacerbated divisions within OPEC. The oil companies, for obvious reasons, exploited this lack of cohesion and used various "shenanigans" to promote disunity and discord within the ranks of OPEC. They frequently used delaying tactics and created impasses emanating from the need for procedural clarification. Other tactics popular with the oil companies were making subtle inducing offers to individual members in an attempt to undermine OPEC's collective bargaining efforts; deliberately making offers that were bound to be rejected; or making last-minute modifications of previous offers immediately before OPEC conferences.[49] The cumulative intent of these actions was to keep the oil states in disarray and to postpone any decision that would diminish the companies' power and prestige. They ultimately obtained their purpose; OPEC failed to resolve the royalty-expensing issue until 1968.

The nature of OPEC's concern during the period covered in this chapter, and the responses of all actors—the oil states as well as the multinational oil corporations—were largely economic. Even if politics entered into the calculus of OPEC, its impact appears to have been negligible.

3. The Making of the Sellers' Market

Beginning in 1967 and continuing through 1971, the international oil market was transformed from a buyers' to a sellers' market. The uncertainties of the sellers' market complemented the objectives of the oil-exporting countries as excessively as the uncertainties of the buyer's market had complemented those of the oil companies and consuming nations. OPEC members, with Libya serving as a pacesetter, took full advantage of the changed market conditions through three agreements—Tripoli I, Teheran, and Tripoli II—and brought about substantial increases in the prices of their crude oil. The sellers' market resulted from the cumulative impact of a variety of factors. OPEC's ability to negotiate these agreements in the sellers' market of the early 1970s eventually enabled this organization to establish itself as a price-manipulating entity, especially in 1973 and again in 1979 and 1980.

A number of these factors grew out of the Suez crisis. For the previous hundred years, the Suez Canal had served as a vital link between the oil-exporting countries of the Middle East and their entrepreneurs, the Western European states. The closure of this canal in the aftermath of the Arab/Israeli war of 1967 affected the international oil trade in several significant ways. The short-cut route between the Persian Gulf and Western Europe was severed, forcing oil tankers to navigate the Cape of Good Hope to reach their Western European markets. As a result, tanker rates and freight costs increased throughout the world because of the added mileage. Another significant result of the closure of the Suez Canal was that such countries as Libya and Algeria successfully introduced price increases based on locational advantage.[1]

The closure of the Suez Canal affected the Mediterranean and the Persian Gulf oil-producing countries in starkly different ways. Of the Mediterranean states (Libya, Algeria, Saudi Arabia, and Iraq), Libya and Algeria had locational advantages because of their promixity to Western Europe. The Persian

Gulf producers, on the other hand, were now in the precarious situation of having to ship their commodity to Western Europe around the Cape of Good Hope. Even Saudi Arabia and Iraq, with limited access to one Mediterranean outlet (the Trans-Arabian Pipeline, known as Tapline), had to ship the bulk of their oil around the Cape. Because the oil from the Persian Gulf would naturally cost more because of higher freight, the Mediterranean states initiated demands for higher posted prices on the grounds of locational advantage.[2]

The revolutionary government of Libya, which came to power in a bloodless coup on September 1, 1969, was the pacesetter in negotiations between OPEC members and the multinational oil corporations. The negotiating posture of the revolutionary government of Libya could best be described as unorthodox. Under the Sannusi rule, Libyan oil affairs had been handled by decision makers who were totally ignorant of the oil business and were ready and willing to sell substantially favorable concessions to the multinational oil companies for hefty deposits in their Swiss bank accounts.[3] The bitter memories of this experience were fresh when the revolutionary government of Colonel Muammar Qaddafi initiated demands for negotiations with twenty-one oil companies operating in Libya. On January 20, 1970, the Libyan oil minister, Izzedin al-Mabruk, began negotiations on the government's proposal to raise the posted prices of crude in Libya.[4]

Libya's approach to these negotiations was rather elaborate. Publicly, Libya heightened the level of its accusatory rhetoric in order to apply pressure on the oil corporations. In a blistering speech on January 29, Colonel Qaddafi claimed that Libyan oil was priced too low in relation to its production cost, high quality, and proximity to markets. This overture was followed by a series of public and private measures intended to convince the oil companies that the new Libyan rulers meant business. The government sent a series of letters to its concessionaires ordering them to initiate drilling within thirty days on all concessions where such action had not begun. At the same time, the government announced cooperative agreements with Algeria and other Arab oil states. Soviet oil experts were invited to inspect the Libyan oil fields.[5] On February 27, in a communiqué issued at the conclusion of the visit of Yugoslavian President Josip Tito, Libya announced its intention to cooperate with Yugoslavia on such varied oil-related matters as the building of an ammonia plant, the development of a tanker fleet, and the training of oil technicians.

Such contacts served two important purposes for Libya, one substantive and the other symbolic. Substantively, these contacts utilized the technological know-how of these countries to develop indigenous expertise in the upstream operations; symbolically, such contacts signaled the Western oil industry that, in the wake of their refusal to accede to Libyan demands for price increases, the government might not hesitate to nationalize their assets and switch to other equally competent avenues. As an additional demonstration of its resolve to accumulate technological knowledge, Libya in March 1970 established the Libyan National Oil Corporation (Linoco), the mandate of which included joint ventures and other downstream activities. Although the message underlying these Libyan maneuvers was only too clear to the oil companies, they were not willing to capitulate easily because doing so was bound to establish a dangerous precedent which would encourage other oil states to emulate Libya.

Libya had a number of independent oil companies and several international majors operating within its borders. The Libyan rulers were cognizant that they could not apply pressure on such oil giants as Exxon, but they believed that the independents, because of their inordinate dependence on Libyan oil, might succumb if they applied the right kind of pressure. This "right" kind of pressure turned out to be the government's decision to reduce the oil supplies of the independents. Using conservation as the rationale, the Libyan government decided to curtail the production of its concessionaires, especially that of the independents. On May 27, 1970, Mabruk announced that the production rates of Occidental (the largest independent producer, which depended on Libya for half its crude oil supplies) from Intisar oil field had been reduced from 320,000 barrels per day (b/d) to 200,000 b/d for purposes of conservation. On June 12, Occidental's total production of Libyan crude, including that covered by the May 27 reduction, was reduced from 800,000 to 500,000 b/d. Amoseas (a joint venture between Texaco and Standard Oil of California) was ordered on June 19 to reduce its production by 100,000 b/d, a reduction of about 26 percent from its May production of 384,000 b/d. On July 9, Oasis (a combination of the Continental Oil Company, the Amerada Hess Corporation, and the Royal Dutch-Shell Group) was ordered to reduce production by 150,000 b/d, to 895,000 b/d. On August 4, Esso was also ordered to cut its production, from 740,000 to 630,000 b/d; this cutback, however, was not to be initiated until September 5.

In addition to production cutbacks, the Libyan government kept its concessionaires from developing a unified negotiating proposal by refusing to negotiate with them as a block, adopting instead a strategy of separate negotiations. When these parlays failed to produce results to the government's liking, the director general of Technical Affairs in the Oil Ministry announced that Libya would take unspecified unilateral actions if the oil companies did not agree to raise prices and would coordinate its policies with Algeria, which was also seeking an increase in its crude price from $2.08 to $2.65/b.[6]

Libya was fully aware of its own strength, which was manifested not only in the growing demand for Libyan crude oil but also in the continuing overall growth of oil consumption in the non-Communist countries. In 1969, more than 25 percent of Western European oil needs were met by importing Libyan oil. Concurrently with this growing dependence on Libyan crude, coal production in

Western Europe was falling off at higher rates than projected. As a result, the oil companies were under increasing pressure from their customers to raise their levels of supply. In May 1970, when the Tapline was accidentally ruptured, an equivalent of 480,000 b/d of Saudi oil stopped flowing, and the Syrian government assured Libya that they would not allow the reopening of the Tapline before the end of the summer. Such assurances were based on the fact that Syria was benefiting from the closure, since it had negotiated new long-term sales of 3.5 million tons of oil each year at $1.50/b instead of $1.28/b, the price Syria had secured for previous sales. In addition, according to Marwan Iskandar, Syrian leaders could not fail to remember that the Saudi decision to close the Tapline for two months following the war of June 1967 had forced Aramco "to concede full expensing of royalties on Mediterranean deliveries," a concession that enhanced Saudi Arabia's income by $21 million a year. Since this incident, Syria had been looking for avenues to double its transit royalties from the Tapline, but attempts had been unsuccessful. The accident of May 1970 could not have taken place, from the Syrian perspective, at a more appropriate time.[7]

As a result of the Libyan maneuvers, in conjunction with the loss of Saudi oil from the continued closure of the Tapline, the supply situation in the international oil market was becoming critical. Between May, when the conservation policy was implemented, and September of 1970, Libyan oil production was reduced from 3.6 million to 2.8 million b/d.[8] Under such circumstances the oil companies were bound to accede to the government's demand for an increase in the price of Libyan oil. Libya thus performed for OPEC the ground-breaking task of manipulating the rapidly emerging sellers' market by concluding a major agreement, Tripoli I, with the oil companies operating within its borders. This agreement will be discussed shortly.

Another factor which substantially strengthened the bargaining position of the oil-producing countries was the change in supply and demand in the international oil market. At the time of OPEC's formation, the oil market remained a buyers' market. A variety of factors that emerged in the late 1960s transformed the oil market to a sellers' market.

On the demand side, the world energy consumption picture (including coal, oil, natural gas, and primary electricity) was characterized between 1950 and 1970 by sustained increases. Expressed in units of 10^{15} BTU's, world consumption increased from 76.8 in 1950 to 124.0 in 1960 to 214.5 in 1970. For the years 1971 and 1972, aggregate world energy consumption was reported to be 223.5 and 237.2 10^{15} BTU's.[9] Table 2 provides a meaningful picture of the sharp increases in world consumption of different energy sources. Two significant sources of energy, coal and oil, were characterized by sharp alterations in patterns of consumption. Coal consumption, which was 55.7 percent in 1950, declined to 31.2 percent in 1970; oil consumption, in contrast, increased from

Table 2. Sources of World Energy Consumption; 1950–1972 (percent)

Source	1950	1960	1965	1968	1970	1971	1972
Coal	55.7	44.2	39.0	33.8	31.2	29.9	28.7
Oil	28.9	35.8	39.4	42.9	44.5	45.2	46.0
Natural gas	8.9	13.5	15.5	16.8	17.8	18.3	18.4
Primary electricity	6.5	6.4	6.2	6.5	6.5	6.6	6.9

Source: Joel Darmstader and Sam Schurr, "The World Energy Outlook to the Mid-1980's: The Effects of an Alternative Supply Path in the United States," *Philosophical Transaction of the Royal Society of London,* Series A, May 1974, pp. 413-30.

Table 3. Energy Consumption in Selected World Regions, 1955–1980 (in million metric tons oil equivalent)

	1955	1960	1965	1966	1975[a]	1980[a]
United States	857	969	1,194	1,260	1,683	2,000
Western Europe	484	555	697	723	1,031	1,255
Japan	44	73	117	129	249	365
Asia[b]	49	70	99	103	217	322
Latin America	65	92	123	129	250	359
World	2,147	2,822	3,488	3,673	5,752	7,464

Source: Sam H. Schurr and Paul T. Homan, *Middle Eastern Oil and the Western World: Prospects and Problems* (New York: American Elsevier Company, 1971), p. 173.

[a]Projections.

[b]Excludes Japan and the Middle East.

28.9 percent in 1950 to 44.5 percent in 1970. The sharp increases in energy demands are evident also in Table 3, showing energy consumption in various regions of the world. In every instance energy demand was on the rise, and from every indication it would remain on the rise until 1980, the last year for which energy projections were available at that time.

For a variety of reasons the supply situation in the international oil market for the 1970s was becoming increasingly inelastic; that is, price increases were not leading to increases in the availability of supplies of crude. The government of Venezuela had adopted a conservation policy based on the following rationale. First, the government held the view that Venezuela had become too dependent on oil and that this dependence was dangerous to the nation's well-being in the long term. Therefore, it took deliberate steps to reduce the oil industry's share of the total economic output by encouraging diversification into other fields. Second, the government felt that expanded production and sale were at the expense of the price obtained for each barrel of oil. Instead, it opted to maintain price levels by restricting output, thereby conserving oil for future economic development.[10]

The Venezuelan conservation policy, however, was not part of a pattern involving other oil-producing countries. The Middle Eastern countries remained most interested in expanding their rate of production, a venture which deserves a closer look. Table 4 indicates a sustained increase in the absolute rate of production, but the figures on percentage changes in production present a slightly different picture. Iran was the only Middle Eastern country to increase its share, from 15.7 percent in 1956 to 18.1 percent in 1966. The other three Middle Eastern states, Iraq, Saudi Arabia, and Kuwait, declined relatively. The most significant change in comparative terms was the increase experienced by Libya, from very little production until 1961 to almost 13 percent of the combined Middle Eastern and North African total in 1966 and further relative increases in 1967 and 1968.

The Middle East was destined to play a crucial role in meeting the growing demand for oil. It remained the principal supplier for Europe and Japan for some time and was rapidly becoming the major source for the United States. Western Europe consumed close to half of the total Middle Eastern exports of approximately 10.5 million b/d and almost the entire export of over 3.5 million b/d from North Africa. Put differently, Western Europe's share of consumption in 1968 averaged about 60 percent of the combined exports from the Middle East and North Africa. According to projections, the overall high share of exports from these regions was expected to remain unchanged by 1980. Japan consumed about 25 percent of the Middle Eastern exports in 1968 and that figure was projected to increase to nearly 30 percent by 1980.[11]

A closer and more elaborate look at the increased dependence of the United States on Middle East oil will demonstrate that until the early 1960s most of the American oil demand was met by domestic production. Within the next ten years, however, America imported a third of its oil.[12] In his testimony before the Senate Foreign Relations Committee on May 31, 1973, William E. Simon, then deputy secretary of the Treasury, stated that domestic demand for oil in the United States had increased from 15.1 million b/d in 1971 to approximately 18 million b/d in 1973 and was expected to jump to about 21 million b/d in 1975 and to about 25 million b/d by 1980. He also added that the U.S. dependence on foreign oil was expected to increase from 27 percent of the total consumption in 1972 to about 33 percent in 1973 and over 50 percent in 1980.[13] In other testimony, Simon said that the United States had to increase imports of foreign

Table 4. Middle Eastern and North African Oil Production, 1956–1968

Year	Iran	Iraq	Kuwait	Saudi Arabia	Neutral Zone[a]	Other Middle East[b]	Total Middle East	Algeria (incl. Sahara)	Libya	Other North Africa	Total North Africa	Total Middle East and North Africa
						Thousand b/d						
1968	2,835	1,505	2,420	2,830	425	1,170	11,185	920	2,600	295	3,815	15,000
1967	2,595	1,225	2,290	2,600	420	830	9,960	840	1,745	170	2,755	12,715
1966	2,110	1,390	2,275	2,395	420	715	9,305	715	1,505	130	2,350	11,655
1965	1,905	1,315	2,170	2,025	355	570	8,340	560	1,220	125	1,905	10,245
1964	1,705	1,255	2,115	1,730	360	450	7,615	560	860	125	1,545	9,160
1963	1,490	1,160	1,930	1,630	315	295	6,820	505	465	110	1,080	7,900
1962	1,330	995	1,830	1,525	245	250	6,175	435	185	90	710	6,885
1961	1,195	990	1,645	1,390	175	220	5,615	330	20	75	425	6,040
1960	1,060	955	1,620	1,245	135	220	5,235	180		65	245	5,480
1959	940	850	1,380	1,100	120	210	4,600	25		65	90	4,690
1958	825	725	1,395	1,005	80	215	4,245	10		60	70	4,315
1957	730	445	1,140	985	65	175	3,540			45	45	3,585
1956	540	630	1,085	975	30	155	3,415			35	35	3,450
						Percentages						
1968	18.9	10.0	16.1	18.9	2.8	7.8	74.6	6.1	17.3	2.0	25.4	100.0
1967	20.4	9.6	18.0	20.4	3.3	6.5	78.3	6.6	13.7	1.3	21.7	100.0
1966	18.1	11.9	19.5	20.5	3.6	6.1	79.8	6.1	12.9	1.1	20.2	100.0
1965	18.6	12.8	21.2	19.8	3.5	5.6	81.4	5.5	11.9	1.2	18.6	100.0
1964	18.6	13.7	23.1	18.9	3.9	4.9	83.1	6.1	9.4	1.4	16.9	100.0
1963	18.9	14.7	24.4	20.6	4.0	3.7	86.3	6.4	5.9	1.4	13.7	100.0
1962	19.3	14.5	26.6	22.1	3.6	3.6	89.7	6.3	2.7	1.3	10.3	100.0
1961	19.8	16.4	27.2	23.0	2.9	3.6	93.0	5.5	0.3	1.2	7.0	100.0

Year	Iran	Iraq	Kuwait	Saudi Arabia	Neutral Zone[a]	Other Middle East[b]	Total Middle East	Algeria (incl. Sahara)	Libya	Other North Africa	Total North Africa	Total Middle East and North Africa
					Percentages							
1960	19.3	17.4	29.6	22.7	2.5	4.0	95.5	3.3		1.2	4.5	100.0
1959	20.0	18.1	29.4	23.5	2.6	4.5	98.1	0.5		1.4	1.9	100.0
1958	19.1	16.8	32.3	23.3	1.9	5.0	98.4	0.2		1.4	1.6	100.0
1957	20.4	12.4	31.8	27.5	1.8	4.9	98.7			1.3	1.3	100.0
1956	15.7	18.3	31.4	28.3	0.9	4.5	99.0			1.0	1.0	100.0
					Increase, 1965–1966							
Thousand b/d	1,570	760	1,190	1,420	390	560	5,890	715[c]	1,505[d]	95	2,315[e]	8,205
% of increase	19.1	9.3	14.5	17.5	4.8	6.8	71.8	8.7[c]	18.3[d]	1.2	28.2[e]	100.0

Source: British Petroleum Co., Ltd., *Statistical Review of the World Oil Industry*, 1964, 1968, as cited in Sam H. Schurr and Paul T. Homan, *Middle Eastern Oil and the Western World* (New York: American Elsevier Co., 1971), p. 70.

[a]Neutral Zone has an ambiguous political status. Insofar as oil is concerned, it is controlled partly by Saudi Arabia and partly by Kuwait.

[b]In recent years, mainly Abu Dhabi and Qatar.

[c]Between 1957 and 1966.

[d]Between 1960 and 1966.

[e]Essentially, between 1960 and 1966.

Figure 2. World Crude Oil Spare Capacity and Demand, 1955-1975 (Excluding Communist countries)

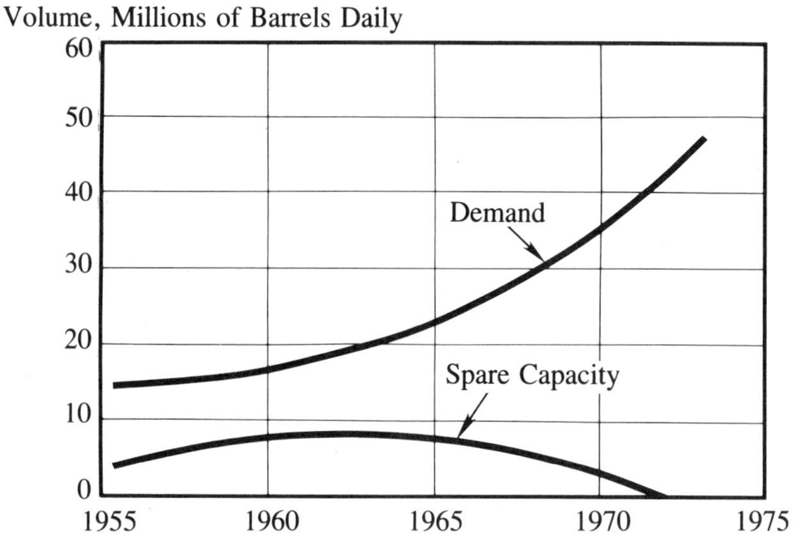

Source: U.S. Congress, Senate, Subcommittee of the Committee on Foreign Relations, Hearings, *Multinational Corporations and United States Foreign Policy*, 93 Cong., 1 and 2 sess. (Washington, D.C.: GPO, 1974), pt. 5, p. 212

oil substantially, particularly from the Middle East. He added that between 1969 and 1972, total imports had risen 52 percent, to 4,685,000 b/d.[14]

The magnitude of structural changes in the overall demand and supply situation becomes obvious with the three-fold increase in worldwide demand for oil, from 15 million b/d in 1955 to 45 million by 1973 (Figure 2). Another measure of rapidly increasing demand was the pace of decline in spare capacity (i.e., idle crude-producing capacity that is available to escalate production under increased demand). As shown in Figure 2, world spare capacity, which stayed around 6 million b/d in the early 1960s, sharply declined around 1970.[15]

The Price Negotiations

The Tripoli I Agreement A tactical maneuver which became the hallmark of the Libyan negotiating posture was the government's refusal to bargain with the oil companies as a united front. Instead, Libya picked its own targets: Occidental, Continental Oil Co., Amerada Hess Corporation of the Oasis group, and other independents. The independent concessionaires, in most cases, were substantially dependent on Libyan oil to stay in business, and the Libyan government was fully conscious of their vulnerability. These companies were chosen to bear the brunt of the government's strong-arm tactics in order to create consternation and uncertainty within the rank and file of the oil companies. The government also continued to harass the oil companies by threatening to expropriate their property or to legislate unilateral price increases.

In the early summer of 1970, the government of Algeria announced a unilateral increase of 70¢/b in posted prices. Algeria justified this upward move in the price of its crude on the basis of locational advantage. In May 1970, Algeria, Iraq, and Libya issued a joint communiqué declaring that they would form a united front against the oil companies and would cooperate in joint oil ventures.[16] By August of that year, as a result of Libyan pressure tactics, Linoco had signed a multitude of joint ventures, the Libyan government was visibly and actively cooperating with Algeria and Iraq, production cutbacks were widely in place, and the oil companies were regularly being given ultimatums threatening more stringent actions. But the concrete evidence of the potency of the Libyan strategy of separate negotiations surfaced with reports that Esso and Occidental were making a variety of offers. On September 4, 1970, Colonel Qaddafi announced that Occidental had signed an agreement with the government, generally accepting the Libyan demands. As a result of this, the Tripoli I agreement, the posted price of crude was raised from $2.23 to $2.53/b, and by 1975 it was to be raised to $2.63/b. These increases also enhanced Libya's take from $1.10 to $1.27/b, and by 1975 the government was to receive $1.33/b.[17]

As the number of companies reaching settlement with the Libyan government swelled, the recalcitrants were left with virtually no enviable choice. Their continued refusal to accede to the government's demand, as they perceived it, might risk a shutdown at a time when such an action would certainly transform a tight supply situation in Europe into an acute shortage. Thus, the remaining members of the industry acceded to the new terms in October 1976.[18]

Even though the Tripoli I agreement was concluded outside the framework of OPEC, the organization remained highly supportive of the Libyan and Algerian endeavors to raise the posted prices of their crude based on locational advantages. In an earlier resolution, passed in September 1967, OPEC had endorsed the Libyan demand for upward price adjustment.[19] OPEC adopted a similar attitude in 1970 by offering full support to Algeria for taking measures aimed at increasing the price of its crude because of its lightness and proximity to European markets.[20] This expression of support, in addition to providing a psychological impetus for Libya and Algeria, also bolstered their bargaining position vis-à-vis the oil industry.

According to Taki Rifai, an Arab analyst of oil affairs, the government of

Libya felt that the oil companies adopted a "negative" negotiating stance from the outset by stating that the levels of Libyan postings were "fair and satisfactory" and that there was no valid justification for a price increase. Such a disposition was reported to have convinced Libya that price negotiations could not be conducted on "rational professional grounds."[21] Rifai's observation concerning the significance of the Tripoli I agreement merits quotation, since it may also portray the reality of the international oil market as perceived by the oil states:

> The precedent-shattering hikes in prices and tax rates were fascinating not only because of their magnitude but also and primarily because of their political significance since they demonstrated that the major oil companies were no longer invulnerable and that they no longer formed a power center independent of the governments of the countries in which they operate. . . . In fact, the thunder of the Libyan settlement started its ball-rolling effect throughout the oil-producing countries before economists and observers of the industry had enough time to evaluate the scope and consequences of the arrangement and to foresee its far-reaching impact.[22]

The Teheran Agreement The success of Tripoli I sowed the seeds of a series of future price agreements. The Teheran agreement was to be the second one. Two events preceding the Teheran negotiations are worth mentioning since they point to a variety of actions and negotiations culminating in the demise of the oil companies as the negotiating partners of the oil-exporting countries. First, in November 1970 the government of Iran initiated a series of negotiations with member companies of the Iranian Consortium. On November 14 an agreement was reached that raised the posted price of 31° heavy oil by 9¢/b and increased income tax rates from 50 to 55 percent. Iran accepted this agreement on an *ad hoc* basis, pending OPEC's forthcoming Twenty-First Conference.[23] Next, the Venezuelans decided not only to raise the income tax on oil company profits from 52 to 60 percent but also to make the increase retroactive to the first of the year. This action was in response to huge deficits confronting the government.[24] These two events were indicators that the iron-fisted grip of the international majors on oil prices was being increasingly loosened.

It is quite clear that the emerging uncertainties of the sellers' market were beginning to show their impact on the erstwhile advantageous position of the oil companies. With the Tripoli I agreement and the aforementioned Iranian and Venezuelan actions serving as a backdrop, the Twenty-First Conference of OPEC convened in Caracas to take, as it turned out, several important actions. One such action was passage of Resolution XXI.120, which was perhaps the single best indicator of OPEC's behavior in the coming years. In this resolution, among other things, the oil states expressed their resolve to initiate a unified pricing strategy. The framework of this strategy was "regionalization," or a regional cooperative approach. The principle involved was to treat separately the price problems of each region within OPEC jurisdiction. The negotiated increases in the posted prices in one region would serve as the basis for ensuing negotiations aimed at incorporating these increases in the other regions. For example, any agreement between the oil industry and the Gulf states to raise the prices of crude would become the basis of renewed negotiations for more price hikes by the Mediterranean states, and vice versa. Three regional committees were established to deal with the price issue: the Gulf states committee, made up of Iran, Iraq, and Saudi Arabia; the Mediterranean committee, including Libya, Algeria, Iraq, and Saudi Arabia; and a third committee comprised of Venezuela and Indonesia.[25]

The Gulf states committee was to initiate the Teheran negotiations on a number of well-defined demands formulated by OPEC.[26] First, the committee was to propose a new, higher posted price for 40° API crude, which was to serve as a benchmark for all Gulf crudes. The prices of all other crudes were to be raised accordingly with the proviso that the deescalation per degree of gravity below 40° API be fixed at 1.5 cents instead of the 2 cents agreed to under Tripoli I. The Gulf states were also to encourage replacement of the prevailing gravity system that regarded any crude between 34.0° and 34.9° API as 34° API. Instead, the governments involved wanted to regard any crude above 34° as 35°, and so on within each degree of gravity. Second, in an attempt to equalize the rate of taxes agreed to in Tripoli I, the Gulf states committee was to press for an increase in income taxes to a minimum of 55 percent. Third, the Gulf committee was to demand elimination of the OPEC allowances, an act that would give the governments an additional income of about 4¢/b on 34° API for 1971.[27]

The Caracas conference also affirmed OPEC's united stand against potential attempts by the oil companies to employ discriminatory production policies to penalize any member state for actions taken in pursuit of OPEC objectives. The conference, without incorporating it in a formal resolution, also commissioned a study to probe the possibilities of linking posted prices of crude to the value of the U.S. dollar in terms of purchasing power. This attempt was the beginning of long and drawn-out, though unsuccessful, endeavors by OPEC to guard against inflationary trends by indexing oil prices to the value of the dollar in real terms.[28]

Before the beginning of the Teheran negotiations, which were originally scheduled for January 12, 1971, Libya, on January 2, presented the oil companies with a new set of demands nullifying the Tripoli I agreement. The companies felt, according to Under Secretary of State John Irwin II, that acceptance of the new Libyan demands would start a new "chain reaction" and that their only hope "against being whipsawed by successive demands from different producers was to conduct negotiations with OPEC as a whole."[29]

During the Tripoli negotiations the oil companies had failed to present a united front against Libya. In fact, during the most precarious days of Tripoli I, Occidental, which was suffering the worst impact of the Libyan conservation policy, sought Exxon's assistance by proposing a "Safety Net Agreement." According to this agreement, if a company's supplies were reduced or assets were nationalized by Libya as a pressure tactic, other companies operating in that country would share a percentage of their crude with the penalized company. The Libyan operators, in turn, were to be compensated by companies operating in the Persian Gulf countries.[30] When Exxon refused to provide Occidental the needed supplies under this proposed agreement, Occidental had no alternative but to yield to Libyan demands. But Libya, by attempting to renegotiate Tripoli I, was in violation of the age-old practice of "sustenance of an agreement through its duration." The oil companies, again mindful of the long-range ramifications of giving in to Libyan demands to renegotiate Tripoli I, decided to enter the Teheran negotiations as a united front. With this *modus operandi* firmly established, the American multinationals sought clearance from the Antitrust Division of the U.S. Department of Justice. Such clearance was granted on January 13 in the form of "Business Review Letters." On the same day the oil companies sent a "Message to OPEC" calling for joint negotiations.[31] By issuing this invitation, they hoped to pit the moderate oil states, such as Saudi Arabia, Kuwait, and Iran, against the price hawks, Libya and Algeria. The companies were aware that the hardline posture of these two states would keep them from being a party to any agreement that was designed to apply to all members and had long-term validity.[32]

OPEC turned down this invitation to start what it perceived as an all-or-nothing bargaining approach. Instead, it decided to play a "consultative role," according to J.E. Hartshorn, an expert on international oil affairs. The member states were given the responsibility of conducting negotiations with their concessionaires, and OPEC remained in charge of advising on the details of suitable demands, offering "a forum to share militancy and compare notes," and passing "resolutions committing all member governments to support each other."[33]

The United States government, which had maintained a policy of noninterference during the hectic days of Tripoli I, decided to play an active role just prior to the Teheran negotiations. The State Department established an interagency task force and began consulting with England, France, and the Netherlands, whose companies were also to become parties to the Teheran negotiations. The State Department also endorsed the idea of the oil industry negotiating with the OPEC countries as a bloc instead of separately. Under Secretary of State Irwin was sent with personal messages from President Nixon to each of the three leaders of the Gulf states: Iran, Saudi Arabia, and Kuwait. The purpose of the Irwin mission was to explain the reasons underlying the antitrust waivers granted the oil companies, to seek assurances from the Gulf producers that they would continue to supply oil at reasonable prices to the free world, and to avoid a harmful impasse between the oil states and the oil companies. According to Irwin, the three moderate Gulf states firmly rejected joint negotiating offers made by the oil companies. They did, however, express their willingness to enter into a five-year agreement with the companies and to abide by the agreement even if increases were to be negotiated in other areas, an apparent reference to the renewed Libyan efforts to renegotiate the Tripoli I agreement. On the basis of the findings of the Irwin mission, the State Department recommended to the companies that they open negotiations with the Gulf producers and conduct parallel negotiations with Libya.[34]

The organizational center of the oil companies, otherwise known as the London Policy Group, was comprised of Lord Strathalmond, managing director of British Petroleum; George T. Piercy, senior vice president and director of New Jersey Standard (Exxon); William P. Tavoulareas, president of Mobil; Alfred DeCrane, vice president of Texaco for the Eastern hemisphere; and John E. Kirchner, president of Continental Oil for the Eastern hemisphere division. The OPEC negotiating team was comprised of Dr. Jamshid Amouzegar, the finance minister of Iran, as chairman; Ahmad Zaki Yamani, the minister of petroleum and mineral resources of Saudi Arabia; and Saadoun Hammadi, the minister of oil of Iraq. This group was frequently joined by Abdul Rahman Saleem Atiqi, the finance and oil minister of Kuwait.

When the representatives of the two sides gathered in Teheran, the oil states were clearly the advantaged actors and the oil companies were behaving as actors whose previously advantaged position was fast eroding. The oil producers were expected to get what they wanted in the short term, including protection against inflation; and the oil industry was expected to act in such a way as to minimize losses as much as possible. It was also becoming apparent that the conclusion of the Teheran negotiations to the satisfaction of the oil states was likely to serve as a prelude to another series of demands triggering further negotiations.

When the negotiations began in Teheran, it was quite apparent to the oil companies that separate negotiations, one with the Gulf states and the other with Libya, could lead to further leapfrogging in the prices of crude. The companies were also aware that their insistence on a single negotiation might seriously risk an impasse, a situation which, in turn, might lead to imposition of a system wherein prices would be unilaterally established by the producing governments. In order to safeguard against leapfrogging, the oil companies sought assurances that the Gulf countries would not reopen the agreement, no matter what transpired in Libya, and would not use production cutbacks to help Libya should that country impose exorbitant demands on the companies. The Gulf countries provided the companies with such assurances.[35]

The OPEC members set a definite deadline, February 3, for the successful conclusion of negotiations. The talks ceased on February 1, however, because the companies considered the demands of the producing countries too drastic and the deadline too stringent. The deadline was extended as a result of an Extraordinary Meeting of OPEC in Teheran, where, after hearing a progress report, members decided that the Gulf states were to take the necessary legal and legislative measures to implement the objectives embodied in Resolution XXI.120 on February 15. If the concerned companies failed to comply with these measures within seven days of their adoption, member states were to take appropriate actions, including total embargo on shipment of crude oil and petroleum products by such companies.[36]

Throughout the negotiating period, the threat of nationalization hung like the sword of Damocles over the heads of the oil company representatives. The Shah of Iran maintained pressure on the oil industry by a combination of carrot and stick diplomacy. While using the threat of nationalization as the stick, he induced cooperation by assuring the companies that the Gulf states would honor a five-year contract even if other oil nations leapfrogged the settlement.

The oil companies, for their part, sought to soften the negotiating posture of Iran and Saudi Arabia by poorly disguised threats to link oil negotiations to the security of the Persian Gulf. It was conveyed to these two oil states that excessive harshness toward the companies at this time, while the producers had them over a barrel, might make Western Europe so unhappy that even the United States and Britain might be reluctant to pay for a military presence in the Gulf and Mediterranean of sufficient size to keep the Russians out. The oil industry exerted every effort to make sure Iran and Saudi Arabia clearly understood this message. The *Economist* characterized the activities of the oil industry as "a gunboat diplomacy of a sort, but played with a butterfly net."[37]

The already-shaky bargaining position of the oil companies was further weakened by a background paper circulated during the negotiations which pointed out that there was no escape from dependence on the Middle East. According to this paper, of the billion tons a year consumed in the non-Communist world, the United States was utilizing 700 million. One-fifth of the American needs were met by imports, and this market was growing by 25 million tons a year. So even if Alaska's North Slope reached 100 million tons a year by 1975, the American imports would be 125 million tons. Consequently, the paper warned, even the fabulous finds of the North Slope could do no more than blunt the edge of the growing need for Middle East oil.[38]

On February 14, one day before the deadline set by OPEC members, the final agreement was signed. The Teheran agreement included: (1) An immediate increase of 35¢/b in the posted price of Gulf crudes, of which 33 cents was in the form of "uniform" increase and 2 cents was in settlement of freight disparities. (This increase was applicable to 40° API crude. The price of 31° API

Table 5. Additional Revenues to Gulf Producing Countries for 1971 (Resulting from Settlements of November 1970 and Teheran Price Agreement of February 14, 1971) (*MEES* Estimates, Major Concessions Only)

Country (Major Concession[s])	Est. Exports, 1971 at Gulf Terminals (million barrels)	Additional 1971 Revenue ($ million)		
		Nov. 1970 Ags.[a]	Feb. 1971 Ags.[b]	Total
Iran (Consortium)	1,350	120	380	500
Saudi Arabia (Aramco)	1,300	110	364	474
Kuwait (KOC)	1,058	125	290	415
Abu Dhabi (ADPC/ADMA)	288	20	78	98
Iraq (BPC)[c]	210	14	64	78
Qatar (QPC/Shell)	146	10	42	52
Total	4,352	399	1,218	1,617

Source: *Middle East Economic Survey,* Feb. 19, 1971.

[a]Increase in tax rate from 50% to 55%; increase of 9¢/b in posted prices of heavy and medium crudes.

[b]Increase of 35-40¢/b in posted prices; elimination of OPEC allowances.

[c]Calculated as if terms were uniform with those of other Gulf producers.

crude was to be increased by 40.5¢/b; the prices of crudes less than 30° API [mainly high-sulphur oil from Saudi Arabia and Neutral Zone] were to be settled at a later date.) (2) An annual increase of 5¢/b as a flat-rate settlement in the prices of the refined product that the companies might introduce in the coming years, and an annual increase of 2.5 percent in posted prices to compensate for the erosion of the purchasing power of oil states as a result of worldwide inflation. (3) A willingness on the part of oil companies to eliminate the OPEC allowances which increased the income of the oil states by 3-4¢/b for 1971. (4) A stabilization of tax rates at 55 percent. And (5) A willingness on the part of the oil states not to seek any further increases in the terms of settlement for a period of five years.[39]

The gains of the oil companies as a result of the Teheran agreement included security of supplies, a guarantee against price leapfrogging, and no threats of embargo from February 15, 1971, through December 31, 1975.[40] The Teheran agreement was clearly an enormous victory for OPEC, especially for Saudi Arabia, Iran, Iraq, Kuwait, Abu Dhabi, and Qatar. Translated into dollar amounts, this victory reveals, as shown in Table 5, that the revenues for five oil states were estimated to at least triple and in one instance to more than double in the course of four months.

The Tripoli II Agreement Even before the beginnings of the Teheran negotiations, the Libyan government expressed its dissatisfaction with the outcome of the Tripoli I agreement. On January 2, 1971, Libya started issuing a series of "nonnegotiable demands" aimed at pressing the oil companies for further price escalations. The Libyan justification for additional increases in the price of crude from the Mediterranean states was similar to the one used to justify Tripoli I; namely, that since the oil companies had agreed to raise posted price profits going to the Gulf states from 50 to 55 percent, Libya would have to ask for an increase of over 55 percent in order to maintain its favorable position resulting from continued closure of the Suez Canal.[41]

The deputy premier of Libya, Major Abdulsalam Jallud, called for implementation of the Caracas resolutions in the following manner:

OPEC Resolutions	**Libyan Implementation Proposals**
1. Resolution XXI.120 called for: (1) increases in the income tax rate in all Gulf states to 55 percent;	The September 1970 increases in tax rate were to be put on top of the new OPEC standard rate of 55 percent, producing new rates varying from 59 to 63 percent.
(2) elimination of posted price disparities by tying price to the highest existing posting and incorporating gravity premia locational surcharges;	
(3) establishment of a uniform increase to reflect general market improvement; (4) a new gravity escalation system; (5) elimination of all OPEC allowances granted to oil companies.	Libya was to receive whatever increases were obtained in the impending Teheran negotiations. In addition, concessionaires were to pay taxes and royalties monthly instead of quarterly as in the past.
2. Resolution XXI.123 supported Libyan efforts to increase exploration.	Libyan concessionaries were to reinvest in that country 25 cents for each barrel of oil exported; non-oil investments would satisfy this requirement.
3. Resolution XXI.124 supported Libyan attempts to get a freight and Suez premium and promised to support "any appropriate measures."	Libya demanded increases in posted prices of 69¢/b to eliminate "excessive windfall profits." Some elements of this were to be temporary and some or all were to be applied retroactively to the closing of the Suez Canal in 1967.[42]

The Libyan rejection of the Tripoli I agreement, which was to have been in operation for five years, was a precedent of enormous significance in oil affairs. The oil industry was not accustomed to such behavior and its natural reaction was to orchestrate an all-out attempt to reestablish conditions which would prohibit any future repetition of the Libyan nullification of a contract.

As discussed above, the oil companies attempted to conduct OPEC-wide negotiations but were forced to abandon the effort. In both instances, the oil companies had the approval and blessings of the U.S. government; it was Libyan rejection of the joint negotiating approach that was the decisive factor behind its abandonment. Prior to initiating the Teheran and Tripoli II negotiations, however, the oil companies, in one last attempt, proposed "separate but necessarily connected negotiations" which would have the semblance of separate negotiations but whose end product would be no different from that of joint negotiations. Essentially this proposal involved negotiations with a group comprised of fewer than all OPEC members; the negotiated settlement, however, had to be accepted by the total OPEC membership and had to guarantee no further leapfrogging. The logistics of this strategy involved the following: (1) The specific proposals would be presented simultaneously in Teheran and Libya. (2) The oil industry negotiating team would split into two halves, each half containing representatives from the major oil companies and Libyan independents. (3) Each half of the team would make it clear that it was only one-half of one team and would endeavor to persuade Libya and the Gulf countries to get together. (4) Neither half of the team would negotiate on the proposals or counterproposals. This was absolutely essential. The line to be taken would be that any counterproposal must be considered by the team in one whole so that the team as a whole could consider what new terms, if any, it was prepared to put forward to both groups.[43] The London Policy Group, convinced of the soundness of their negotiating strategy, selected two negotiating teams, one headed by Lord Strathalmond, the other by George Piercy. Libya, however, dealt a *coup de grâce* to the "separate but necessarily connected" strategy by refusing to negotiate with the industry team headed by Piercy. Instead, Major Jallud received Piercy alone as a representative of Esso. The Libyan rejection of their strategy convinced the oil industry that regional negotiations were a *fait accompli*.

In the absence of any negotiating leverage provided by the industrial consuming countries, and in the prevalence of a growing dependence of oil consumers on OPEC oil, the industry remained quite vulnerable to the coaxing of OPEC. It was forced to yield to a number of demands that it had traditionally deemed "nonnegotiable." The sustained weakening of the oil companies, almost in the manner of a zero-sum game, enhanced the capability of the oil states to ignore the principle of sanctity of contract by renegotiating a recently negotiated agreement. In this losing battle, the oil industry's last stand was to extract a "commitment" from the Gulf states whereby they would refrain from further leapfrogging if the Mediterranean states concluded a better deal. The desperation of the oil companies became quite apparent in a message sent by the London Policy Group to the Gulf negotiating team prior to the Teheran negotiations: "At the moment we are apart. Apparently the financial terms offered by us are not sufficient and we do not at present have authority to increase these. You tell us that you do not have authority to settle the East Mediterranean position or to give assurances which will prevent the leapfrogging from the Mediterranean into the Gulf, and leapfrogging within the Mediterranean. Will you please go back to OPEC and get authority to negotiate for your oil in the East Mediterranean and to widen assurances to cover the points we need covering?"[44]

When the negotiating parties initialed the Teheran agreement, two things became abundantly clear to the oil industry: that in the sellers' market there was little they could do to enforce any future negotiated settlement, and that it would not be long before the Gulf producers would follow the Libyan example of nullifying an agreement under the pretext of "changed environment" or the "privilege [prerogative] of a sovereign." On paper the Teheran agreement promised stability of prices and sustenance of agreement for five years. But the fact that the oil industry was gearing up to renegotiate Tripoli I with Libya and the fact that the two major oil producers who had signed the Teheran agreement, Saudi Arabia and Iraq, were going to participate as members of the Mediterranean team underscored the inherently temporary nature of the Teheran agreement.

Prior to the beginning of Tripoli II, the Libyan bargaining position was formidable for a variety of reasons. First, it was clear that Libya would once again concentrate on independent concessionaires whose dependence on Libyan oil remained as acute as it had been during the Tripoli I negotiations. Second, Libya was once again coordinating its policy with Algeria, its proven ally on oil matters. Third, Libyan foreign exchange reserves were reported to be $1.4 billion. With its low import bill, Libya alone among OPEC members could cease oil shipments for months without suffering economic damage. Finally, Libya continued to enjoy a locational advantage. Its contribution to the world oil export stood at 17 percent, including 41 percent of West German, 32 percent of Italian, and 30 percent of Spanish oil needs.[45]

The Tripoli II negotiations continued with a meeting between the Mediterranean oil ministers on February 25, 1971. In a joint communiqué issued at the end of this meeting, these ministers authorized Libya to negotiate separately with each company operating in that country. If the companies rejected the "minimum demands" agreed upon by these ministers, they were to face a Mediterranean embargo.[46] The Saudi and Iraqi decision to go along with as drastic a move as an embargo was a clear undermining of good faith and a violation of the assurances they had given the oil companies in their capacity as Gulf oil ministers.

OPEC played the role of a restrainer during these negotiations. Its Twenty-Second Conference, which occurred on February 3, passed two resolutions dealing with the Mediterranean situation. Resolution XXII.131 stated:

> And with respect to Algeria and Libya, the necessary legal and (or) legislative measures for the implementation of the object as embodied in Resolution XXI.120 applicable to them shall be introduced at the convenience of their respective governments. In the event that any oil company operating in these member countries fails to comply, within seven days from the date of their adoption, with the same minimum requirements agreed upon by the member countries bordering the Gulf plus an additional premium reflecting a reasonably justified shorthaul freight advantage for their crude oil exports, member countries . . . shall take appropriate measures including total embargo on the shipment of crude oil and petroleum products by such company.[47]

Resolution XXII.132 was designed to discourage the oil companies from taking any retaliatory measures against Libya. Even though the words "gives full support" were printed in bold letters, Rouhani writes that this resolution reflected limited OPEC support for additional Libyan demands. He notes that OPEC's support was available "only in the event that the companies should refuse to Libya, in fixing the price of its crude, the same terms as they would offer in respect to the Persian Gulf area, plus a reasonable freight differential. And it did not extend to any backing for demands concerning retroactivity or obligatory investments, which were known to be part of Libya's claims."[48]

On February 23, local representatives of oil companies met with Deputy Premier Jallud, who called for a posted price of $3.75/b (the prevailing price was $2.55/b). The companies were given three days to start negotiations and two weeks to finish. The Libyan bargaining position received a considerable boost on February 24 when Algeria nationalized 51 percent of all remaining oil production and 100 percent of all gas production that belonged to French oil interests in Algeria. The tough bargaining continued until March 11, by which time the oil industry offer remained at $3.27/b. On the same day, Syria pledged its support for joint action by hinting at nationalizing the Tapline. On March 12, Nigeria announced its decision to join OPEC and sent a delegation to Libya to coordinate policies vis-à-vis foreign oil companies. By March 13, the oil companies raised their offer to $3.31/b. Libya, however, remained dissatisfied with these offers. On March 15, the Mediterranean team announced its intention to cut off oil by an unspecified date if the oil industry failed to accept the Libyan demands.[49]

The negotiating strategy used by Libya during the Tripoli II talks was similar to that used during Tripoli I. The Libyan government mixed threats of nationalization and production cutbacks with continued application of pressure on individual companies. This strategy was bound to produce results favoring Libya since the oil industry, while facing growing uncertainties related to the availability of crude oil in the Mediterranean and Gulf regions, was also finding it increasingly difficult to satisfy the surging demand from consuming nations.

After five weeks of tough negotiations, a five-year agreement was signed between the Libyan government and the oil companies. The agreement included: (1) an increase of 42¢/b in posted prices; (2) an increase in the tax rate to 55 percent for all companies except Occidental, for which it was increased to 60 percent; (3) a low sulphur increment of 10¢/b; (4) four annual increases of 5¢/b; (5) four annual increments of 2.5 percent for inflation; (6) two temporary freight premiums; and (7) a supplemental payment on exports. "The supplemental payments on exports ranged from 8¢/b to 11¢/b, and varied from company to company." As a result of this agreement, the posted price of Libyan crude oil was raised from $2.55/b to $3.447/b. Libya failed to extract a commitment from its concessionaires for reinvestment funds of 25¢/b. The companies, however, agreed to increase exploration activities, and to make monthly rather than quarterly payments. Of the total increase of 90¢/b, Libya was to receive 65¢/b.[50]

According to Rouhani, the gains made by the oil companies as a result of the Tripoli II agreement were as follows:

(1) After a period of uncertainty which lasted over a year, the companies succeeded in securing financial terms that enabled them to plan ahead "with an assurance as to the limits of their commitments."

(2) "At a considerable price," the companies seemed to have bought a stability lasting for five years. During those five years, as in the case of the Teheran agreement, there was to be no change in the terms of the agreement. The companies seemed to have averted the danger of nationalization and to have consolidated their position as concessionaires by gaining security of supplies.

(3) The companies appeared to have obtained a "quit claim" on retroactive demands in return for the surcharge arrangement.

(4) The "element of flexibility" involved in the "agreed variability of the freight differentials" was expected to benefit the companies in the future.

(5) Even with the price increase resulting from Tripoli II, the profit margin of the oil companies from their Mediterranean crude oil supplies was expected to remain "appreciable" compared to the "much lower level of Gulf prices."[51]

It is apparent that Rouhani is presenting these "gains" in the best possible light. The performance of OPEC members during these negotiations clearly established a pattern whereby the period of uncertainty was far from over. On the contrary, the OPEC members were indeed in the process of bringing an end to the era of certainty. The stability factor referred to by Rouhani was a temporary phenomenon, as was the "appreciable" profit margin received by the companies from their Mediterranean supplies.

The foregoing analysis makes it abundantly clear that the growing uncertainties of a fast-emerging sellers' market were beginning to bring about qualitative changes in the status of the multinational corporations, especially in 1970-1971. This was only the beginning of a period during which the oil states were becoming the advantaged actors while the oil industry was gradually becoming a disadvantaged actor. If the oil industry, in the 1960s and the decades previous to that, was accused by the oil states of gross exploitation and of taking excessive advantage of their backwardness and dependence, it was quite clear by the end of 1971 that as advantaged actors the oil states themselves did not behave as neophytes. Arguing that the oil concessionary regimes were inherently unjust because they had been negotiated under conditions of gross asymmetry between the oil industry and oil-producing countries, the latter started to chip away at the dominant position of the multinational oil firms. The growing uncertainties of a fast emerging sellers' market kept the oil industry from effectively countering Libyan maneuvers.

Either by design or as a result of a historical accident, Libya was a natural leader in the move to intimidate the oil industry. The presence of a large number of independents, whose very survival in the oil business was substantially dependent on the availability of Libyan oil, provided a unique and most potent leverage in the hands of the revolutionary leaders of that country. They utilized this leverage ruthlessly to extract the most favorable concessions from the independents and then applied these concessions to all companies operating in Libya. Libyan leaders also proved to be skillful in the use of such negotiating techniques as disallowing a united front by the oil companies, targeting the most vulnerable concessionaires, bringing gradual quantitative increases in production cutbacks, and threatening to nationalize the assets of the recalcitrant companies. Libyan petrodollar reserves outlasted the absorptive capacities of that country's economy while the demand for sweet Libyan crude was on the rise, especially in Western Europe, where coal production was falling off at higher rates than projected. Consequently, Western European energy needs had to be satisfied more and more by crude and fuel oil imports. The continued closure of the Suez Canal also strengthened the already formidable position of Libya.

Libyan success in railroading the Tripoli I agreement was also the result of a substantial lack of ability and foresight on the part of the oil companies and their failure to present a united front during the Tripoli I negotiations. The international majors probably never envisaged the magnitude of Libyan success and, more important, the implications of that success for future negotiations. The uncertainties of the sellers' market enlarged the implications of Tripoli I beyond the wildest imaginings of even the most radical members of OPEC. Undoubtedly the precedent of Tripoli I created a momentum for the Teheran agreement. Yet the Libyan decision to abrogate Tripoli I nullified the Teheran agreement even before it was concluded. It was remarkably naive of the oil industry to assume that even such "moderate" OPEC members as Iran and Saudi Arabia would be content with the Teheran arrangement when it was virtually certain that Libya would do everything to top the negotiated concessions of that agreement.

The Tripoli I agreement gave birth to a new phenomenon, regional cooperation, or regionalization. The pioneers in use of this phenomenon in oil negotiations were Libya and Algeria, but it was not long before other oil states recognized its value and fully used it to their own advantage. The Teheran agreement established regional cooperation as a more or less permanent phenomenon and a source of financial bonanza for the oil states, facilitating leapfrogging in the prices of crude.

The schizophrenic behavior of Saudi Arabia and Iraq when they negotiated both as part of the "moderate" Gulf negotiating team and as part of the "radical" Mediterranean negotiating team had its own salutary impact on the Libyan negotiating stance. Saudi Arabia and Iraq agreed to abide by the Teheran agreement for their Gulf oil, but, soon after signing the Teheran agreement, did not hesitate to draft with Libya a joint declaration threatening a Mediterranean embargo should the oil companies fail to yield to Libyan demands. It was clear that the process of abrogating agreements was becoming routine.

The attitude of the U.S. government in this period did nothing to stabilize the oil industry, to say the least. From a long-standing tradition, the oil companies, both international majors and independents, negotiated oil agreements with the sovereign states but made sure they did not violate U.S. security interests. What that boiled down to in most cases was serving as secure channels of oil for the United States and its allies. The industry had an excellent track record in this matter. The U.S. government, in turn, had left the oil companies, especially in the area of international oil markets, to conduct their own affairs subject only to the antitrust laws of the United States. The heavy reliance on the oil industry meant that the United States never had an elaborate oil policy. Thus, when the oil industry approached the U.S. government for help, the latter responded not in order to promote its oil policy (which did not exist) but largely to provide a semblance of action. If the purpose of the Irwin mission was to persuade the Persian Gulf states to allow the oil industry to conduct OPEC-wide negotiations, this mission was a miserable failure. In addition, the State Department's advice to the oil companies to conduct separate negotiations—advice based on the joint recommendations of Irwin and the U.S. ambassador to Iran, Douglas MacArthur—took the steam out of the joint negotiating strategy before it was given a fair chance. The Irwin mission also provided a unique opportunity for the oil states to study the policy of the U.S. government concerning the impending negotiations at a time when it would have been better to keep them guessing. This author has not had an opportunity to examine the minutes of negotiations kept by the oil states, but the negotiation minutes of the oil industry and the testimonies of Irwin and other State Department officials provide persuasive evidence that the Irwin mission was both untimely and unwarranted from the vantage point of the oil industry and the Western industrial consuming countries. From the viewpoint of the oil states, however, the Irwin mission was a monumental success. It may have played a crucial role in hardening the attitude of the Gulf states for separate negotiations before the opening of the Teheran negotiations and, more important, in the successful conclusion of the Tripoli II agreement.

The pendulum of oil power was swinging in favor of the oil states. Especially after conclusion of the Teheran agreement, OPEC was recognized by the multinational oil companies as a legitimate representative of the oil-exporting countries. The oil industry had recognized OPEC by issuing to its members a collective invitation for negotiations. As a result of this invitation, according to Rifai: "OPEC obtained its 'title of nobility' and achieved its long-awaited political objective of being officially recognized as a 'valid negotiator' representing and speaking for the community of oil producers. Furthermore, a moral responsibility was conferred on OPEC to guarantee the stability and application of any agreement for a period of five years."[52]

In 1970 and 1971, as a budding advantaged actor, OPEC appeared determined to exploit fully the uncertainties of the sellers' market for use against the oil companies. The latter, in a manner characteristic of disadvantaged actors, remained in a quandary at the conclusion of each agreement about the longevity and terms of the agreement. The following passage adequately sums up the hopes and aspirations of the oil companies during this period: "Now that the OPEC countries have tasted blood, the question is whether the agreements so agonizingly reached at Teheran and Tripoli will survive for their five-year term. Oil companies consider that they paid a high price for stability—much too high a price if it is not forthcoming. . . . The industry hopes that the OPEC countries will be reluctant to repudiate pledges made amid such global publicity."[53]

The attitude of the oil-producing countries was exemplified in the statement made by Yamani at the conclusion of the Teheran agreement. Concerning the stability of the oil prices, he said: "Officially, it (the Teheran Agreement) has to live five years, but a serious disparity [in oil prices paid elsewhere, an apparent reference to impending Tripoli II negotiations] would probably lead to another round of negotiations. That is a big if. We shouldn't create it. But if it comes to the surface we cannot ignore it either."[54] This statement not only summed up the chaotic conditions created by the increasing uncertainties of the sellers' market but also indicated the ominous nature of those uncertainties for the oil-consuming countries in the coming years.

4. The World on a Roller Coaster

In the 1970s OPEC's power and prestige surpassed the wildest imagination of its most optimistic founders. The organization expanded the scope and nature of its activities, and as a result Japan and the industrialized countries of the West became the focus of its price decisions. This chapter focuses on the period from 1971 through 1975, years during which the acute uncertainties of the sellers' market so decisively favored the oil states that, starting in October 1973, OPEC ceased its customary consultations with the oil companies over the price issue. OPEC's newly-acquired ability to introduce unilateral price escalations, however, did not insulate the economies of oil states from the pernicious effects of oil power, and OPEC remained preoccupied with minimizing the erosive impact of the ever-accelerating rate of inflation by raising the prices of crude. Although OPEC's price increases temporarily forestalled erosion of the oil states' buying power, these increases contributed to the rate of inflation, thereby further eroding the buying power of OPEC, which then sought additional price increases. Thus a vicious cycle was created. The recession of 1974-1975 created downward pressure on the prices of crude. The oil states were saved from lowering prices only because some states were able to introduce individual production cutbacks. A series of price negotiations between the members of OPEC and the oil companies will illustrate OPEC's continued but futile attempts to stay ahead of the inflation rampant in the non-Communist economies.

The negotiations over participation, a unique manifestation of OPEC's oil power, were carried out to fulfill a long-cherished dream of the petroleum-exporting countries, that of not only owning and operating their industries in the short run, but also of developing integrated oil companies of their own in the long run, as the international majors had done.

In the period from 1971 through 1975, OPEC initiated direct contacts with the Western industrial countries. Underlying these contacts were two significant, though unrelated, reasons: the desire to exchange oil for technology and weaponry, and the hope of extracting substantial concessions from the industrial consuming nations toward the creation of a New International Economic Order (NIEO). The issue of the North-South dialogue, as it was otherwise known, will be dealt with here only as it affected OPEC and as this organization, in turn, influenced the dynamics of these dialogues.

Between 1971 and 1975, OPEC also experienced recurrence of in-group tensions between the price moderates and the price hawks. The moderates, notably Saudi Arabia, operating on the basis of economic realism, refused to go along with the excessive increases pushed by the hawks, such as Libya, Iraq, Iran, and Algeria. The net result of these tensions was nominal increases in the prices of crude between 1974 and 1978. When measured in constant dollars, however, the real price of oil declined in this period.

The Price Negotiations

At the conclusion of the Tripoli II agreement, the oil industry hoped to maintain a period of price stability. The decisionmakers of the OPEC countries, on the other hand, felt that the advantages negotiated in the Tripoli I, Teheran, and Tripoli II agreements were temporary in nature and could be rescinded. In fact, the oil states were beginning to see the gradual erosion of their income as a result either of what they perceived to be deliberate endeavors by the oil companies and industrial consuming countries or of such economic forces as inflation. An overview of significant OPEC rationale underlying the price negotiations during this period is in order.

During the 1960s the realized prices of crudes remained lower than posted or tax-reference prices. But, after comparing the data on oil trade for 1971 and the first half of 1972, OPEC noted that the oil companies were earning increases in realized prices larger than those in the posted prices. For example, in the Japanese market the net profit realized by the oil companies in late 1971 remained in the neighborhood of 35-40¢/b; in 1972, the net profit increased to about 50¢/b. In the Indonesian market the realized price of Minas-type crude was $2.96/b while the posted price was set at $1.70/b prior to the Teheran agreement. Elaborating on the upsurge in realized prices, Dr. Mohammed Sadiq al-Mahdi, then chief of the Economics Department of OPEC, stated that the tax-paid costs of the Persian Gulf countries as a result of both the 1971 Teheran agreement and the 1972 Geneva agreements (which will be discussed shortly) were estimated to increase by 30-40¢/b and 46-54¢/b, respectively. At the same time, said al-Mahdi, the oil companies were expected to be able to raise product prices by as much as 4-6¢ per American gallon, thus raising their revenue "at least five times, and even more than ten times, greater than the real cost increases involved through the posted price and tax changes."[1]

The issue of taxation on petroleum in industrial countries was also a source of considerable dismay for the oil-producing states. An important thrust of petroleum tax policies in the industrial countries was the concept of the elasticity of demand for crude oil and various oil products and the effects of demand elasticity on the prices of these commodities as well as on the taxation levied thereon. According to this concept, commodities with inelastic demands might be heavily taxed since the higher price would not cause a corresponding drop in demand and the government would stand to increase its revenues. The outcome of such tax policies, as studied by OPEC, is clear in the following observation made by al-Mahdi: "The analysis of the weighted average sales taxes on the three main categories of oil products imposed by six West European countries (Belgium, France, Italy, Netherlands, Sweden and the United Kingdom) during the period October-December 1971 indicates the following: for gasolines, for which there are no substitutes, the average sales tax is $23.00 per barrel; for middle distillates where the demand is less inelastic, the sales tax averages $5.96 per barrel; and for fuel oil, where demand is considered to be rather elastic due to the existence of various substitutes, sales tax is only 70 cents a barrel."[2] In addition, continued al-Mahdi, the industrial consumers, were imposing import tariffs on crude oil and oil products under the pretext of protecting indigenous energy sources. As a result of these practices it became possible for them to collect substantial tax revenues. For the United States, for example, the import tariff was 10.5¢/b; for Japan, crude oil import duties were 32¢/b for crudes with sulfur content of more than 1 percent and 26.3¢/b for crudes of less than 1 percent. According to al-Mahdi's comparison, during 1970 the average sales tax earnings on petroleum products of the aforementioned six Western European countries remained at $7.53/b out of the total average price of $13.14/b, while the oil states' average earnings for the same period remained at about $1/b.

For the oil states, the nagging issue of international inflation, which was consistently eroding their buying power, was a source of major concern. These countries were successful in introducing in the Teheran agreement an escalation clause which set the increase at 2.5 percent per annum for the years 1971 through 1975 to insulate OPEC members' purchasing power from irreparable damages as a result of upward inflationary spirals. But after the Teheran agreement, OPEC members realized that the increase set under the escalation clause was unduly low. To support his argument, al-Mahdi cited data collected by the International Monetary Fund (IMF) on the export price index of industrialized countries for parts of 1971. According to this information, the export index for fourteen industrial countries rose by 5.3 percent in the first quarter of

1971 compared to the same period of 1970. On another occasion, al-Mahdi referred to IMF's report of 1972 to compare the alarming rate of inflation in industrialized countries to the prices of crudes between 1960 and 1970. The latter report indicated that "the annual price average increase for all industrialized countries, as measured by GNP deflators, has been 3.3% for the years 1960-1970 and 4.1% for the years 1965-1970. Moreover, the export price index for fourteen major industrialized countries rose to 115 in 1970 (1963 = 100). However, the crude oil posted prices of the major exporting countries of OPEC remained unchanged at their reduced levels during the period 1960-1970."[3]

In view of the above, OPEC members were convinced that in concluding the three agreements discussed in the last chapter, they had not totally eliminated the chronic inequities that prevailed between their economies and those of the industrial consuming nations or between their status and that of the oil industry. They awaited other opportunities to introduce still further increases in the prices of their commodity. The devaluation of the U.S. dollar in August 1971 could not have come at a more appropriate moment.

The Geneva I Agreement OPEC members had long-standing concerns about the terms of trade between their economies and those of the Western industrial countries, about haphazard fluctuations in the currency rates in the industrial financial markets, and about the harmful effects of inflation on their buying power. The historical "Declaratory Statement of Petroleum Policy in Member Countries" issued in 1968 noted that posted or tax-reference prices "shall be determined by the government and shall move in such a manner as to prevent any deterioration in its relationship [that of posted prices] to the prices of manufactured goods traded internationally." On another occasion, OPEC directed the secretary general to undertake a study with a view to linking posted prices to those of manufactured goods of major industrial countries.[4] During the Teheran negotiations when OPEC members were busy incorporating the escalation clause to safeguard against inflation, the oil companies, in a desperate attempt to retain price stability, offered to tie the oil prices into the broad-based commodity and product index so that adjustments would automatically be made for all monetary and inflationary upheavals. OPEC, however, opted for the rigid 2.5 percent arrangement.

On August 15, 1971, President Nixon, to defend the American dollar from the "intrigues" of the international speculators, announced a suspension of the convertibility of the dollar into gold. This decision created mild shockwaves within the ranks of petroleum-exporting countries since most of them had their tax and revenue payments tied firmly to the dollar and had considerable amounts of dollar holdings. OPEC anticipated a possible devaluation of the dollar and passed, during its Twenty-First conference, a resolution calling for adjustments in posted prices "in case of changes in the parity of monies of major industrialized countries which would have an adverse effect on the purchasing power" of the oil states.[5]

As a result of the Teheran agreement and the two Tripoli agreements, the Gulf states were estimated to have received additional revenues of $1.2 billion in 1971. Additionally, based upon an anticipated 10 percent annual increase in production, their income was expected to rise $3 billion by the year 1975. Libyan oil revenues were estimated to increase by at least $700 million a year. Every 1 percent decrease in the value of the dollar, however, meant a loss of $56.7 million to the signatories of the Teheran and Tripoli agreements on the basis of 1970 revenues alone and around $20 million more on the estimated increases scheduled for 1971.[6] The devaluation also caused a major dilemma for the oil companies which had recently concluded three major agreements and were hoping for five years of price stability. It was clear, however, that this event necessitated long and drawn-out negotiations, covering not only the dollar devaluation but also the terms of trade with the West. In September, OPEC members met in Beirut and decided to explore with the oil companies ways and means to offset any adverse effect on their per-barrel real income.

Libya once again proved to be the pacesetter. The government decreed a 3.57 percent boost in the exchange rate for the Libyan dinar, from $2.80 to $2.90. It also announced that the new rates would be applied to all oil payments due October 30. Other countries followed suit. Indonesia boosted the price of its Minas crude by 49¢/b to compensate for the "de facto" devaluation. Nigeria, which joined OPEC in July 1971, raised its prices on the premise that the price agreement it had made with the oil companies in June 1971 included such an increase in the event of devaluation of the U.S. dollar.[7] Algeria was quite independent in setting its own prices in francs and dinars and was therefore largely unaffected by the dollar crisis. Consequently, the only members affected by this negotiation were the six Gulf states and Libya.

The representatives of the oil companies and petroleum-exporting countries gathered in Vienna in November 1971 to determine whether the host government had a legitimate claim and, if such claim was determined, to establish a formula for compensation. The oil states felt very strongly about the legitimacy of their demands. The oil industry, on the other hand, was of the view that any losses incurred were relieved by the effective revaluation of sterling, the currency in which payments were made to most Gulf countries save Saudi Arabia. The companies further argued that the 2.5 percent escalation clause more than compensated for any losses emanating from devaluation of the dollar.[8]

These negotiations were marked by an extraordinary show of in-group cohesion on both sides. For example, Venezuela announced a new oil policy which, because of three important features, was apropriately labeled a "triple-barreled executive maneuver": (1) The Venezuelan bolivar was revalued upward by more than 2 percent, which in turn devalued the "oil dollar." (2) An average boost of 26¢/b in the tax reference price of crude oil exports and still higher prices in products were announced. Revaluation and the price jump brought the average effective increases for crude oil to 23¢/b as of January 1, 1972. Tax reference prices on crude oil and refined products were to go up anywhere from 21 to 90¢/b. (3) A production export control was decreed which also imposed a quota on foreign producers and exporters. Failure to meet the quota by even a small margin was to trigger higher tax penalties.[9] Undoubtedly, by adopting these policies, Venezuela created a momentum in OPEC's favor.

During the negotiations that began in Geneva on January 11, 1972, OPEC demanded a 12.5 percent price increase. Even though the devaluation of the dollar was only 7.9 percent, the organization based its claim on a statement made by Secretary of the Treasury John Connally in December 1971 that the real devaluation was 11.7 percent. On January 10, 1972, the "Group of 22" (the companies operating in the Middle Eastern oil states), in an offer to OPEC, suggested the use of a quarterly index based on the IMF figures reflecting changes in the cost of imports from the Western countries to the oil states. Using the import patterns of 1970 as a base, any supplemental payments that were made in respect to changes in purchasing power were to be the percentage increase derived from the index, minus the effect of 12.5 percent escalation in the Teheran agreement. Satisfactory arrangements for the future conversion of the U.S. dollar into the currency of payment were also to be agreed upon. In addition, under this proposed plan, adjustment in prices paid by the oil companies to the petroleum-exporting countries were to be made each quarter, and no country was to get less than the 2.5 percent escalation agreed upon at Teheran. This proposal served as the basis of the agreement concluded a few days later.[10]

On January 19, OPEC offered another proposal asking for an 8.57 percent increase in posted prices over and above the 2.5 percent annual increase agreed to at Teheran. An agreement between the oil industry and the petroleum-exporting countries of the Gulf and the Mediterranean regions, except Libya, was reached on January 20. This agreement, labeled "Supplemental" to the Teheran agreement, provided an immediate boost in posted prices of the Persian Gulf states by 8.49 percent. This increase was to be effective from the day of settlement, and there was no provision for retroactivity. Under a special provision of this pact, Iraq and Saudi Arabia were entitled to further adjustments for their Mediterranean exports if Libya should negotiate an additional increase. There was also a provision for future price adjustments in response to alterations in the monetary exchange rates. Such alterations were to move up or down in accordance with a mutually agreed quarterly index that would reflect in percentages the average increases of the currencies of certain industrial countries in comparison to the United States dollar since April 30, 1976. Nine leading industrial countries were involved: Japan, Italy, the Netherlands, Belgium, West Germany, Sweden, Switzerland, Britain, and France. The agreement also guaranteed that the posted prices would not fall lower than they had been under the original terms of the Teheran agreement. The escalation provisions negotiated at Teheran, which provided a 2.5 percent increment in posted prices for inflation, plus 5¢/b annually on January 1 of 1973, 1974, and 1975, were to remain in force. The 2.5 percent annual increment was to be added to the posted price applicable for the first quarter of 1972. For the countries receiving payments in sterling, it was agreed that the conversion of these payments from dollars into sterling was to be calculated at the prevailing commercial rate.[11]

The Geneva II Agreement The Geneva I agreement, which added $700 million to the coffers of OPEC members, failed to satisfy the oil states that they were receiving a fair share of profit from the sale of their commodity in the markets of Western industrial countries. One articulate spokesman for such a point of view was Jahangir Amouzegar, chief of the Iranian Economic Mission, who as late as April 1973 scoffed at suggestions that the oil exporters were receiving the lion's share of the retail oil dollar. He stated that, prior to the Teheran agreement of 1971, out of every dollar of sales that oil producers made at the retail level, the Middle Eastern countries received less than 8 cents in combined royalties and taxes, while the consuming countries received an average share of 47.5 percent of retail oil prices (in some countries as much as 78 percent). Amouzegar also observed that "The average price of a barrel of crude oil (of average quality) paid to the Persian Gulf producers in 1970 was about $1.79 or 17 percent below the $2.18 received by them in 1947. . . . Even at the average posted price of $2.60 per barrel early in 1973, the per-barrel oil income in the Middle East in terms of industrial import prices is lower than it was in 1947. Even at $2.60 a barrel, Middle Eastern oil is still a bargain compared with the U.S. domestic price which is about 50 percent higher."[12]

On February 13, 1973, the U.S. government announced a second devaluation of the dollar by 10 percent and thereby raised the price of gold from $38.00 to $42.22 an ounce. As a result of this action, the petroleum-exporting countries were bound to demand renegotiation of the Geneva I agreement, since the formula for automatic adjustment in posted prices in response to fluctuations in exchange rates, as concluded in that compact, was no longer adequate and the increases in crude oil posted prices thereby produced were substantially less than the 10 percent dollar devaluation or the 11.1 percent appreciation in gold prices.

OPEC members convened a meeting in Beirut on March 22, and a three-man negotiating team was appointed: Oil Ministers A.A. Atiqi of Kuwait, Saadoun Hammadi of Iraq, and Izzedin al-Mabruk of Libya, who served as chairman. As expected, the objective of the OPEC delegation was to renegotiate and amend the Geneva I agreement. OPEC wanted compensation not only for the devaluation but also for the added drop in the value of the dollar on the international money markets, which they claimed to be 10-15 percent. (The Geneva I formula was designed to adjust decreases only up to 5-6 percent.) Originally the conferees stressed, however, that they wanted a technical adjustment for purposes of equity and not any revision of the basic five-year price agreements. The oil industry contended that the mechanisms of Geneva I were designed to adjust prices against changes in all major trading countries, not just against the U.S. dollar, and that any revision would be a breach of contract. In the meantime, Algeria and Indonesia raised the price of their crude. The Indonesian action was an independent one. Algeria, however, intimated that it stood with other OPEC members on the need for price adjustments, even though it was conducting negotiations independently of them.

The April 13 and 14 meetings between representatives of the oil companies and OPEC wre limited to technical discussion. The next round in Vienna on April 23 was discontinued because the oil companies were willing to concede minor adjustments in the implementation of Geneva I in light of the second devaluation of the dollar but were disinclined to renegotiate fundamental changes in this agreement. Primarily because of a predominance of price hawks on the negotiating team, OPEC adopted a hardline position. The OPEC negotiators argued that the price adjustment formula under Geneva I was insufficient to recoup their loss of purchasing power from the second devaluation. Thus they insisted on negotiating a completely new pricing structure.

The April 23 meeting ended with an ultimatum to the oil companies and announcement of an Extraordinary Meeting on May 7. Such conferences, by this time, had become something of a permanent bargaining weapon in the hands of OPEC. Almost every demand for renegotiations followed a pattern: OPEC initiated a new series of demands; the oil companies initially refused to negotiate on the grounds of either breach of previous contract or premature conditions; OPEC responded with simultaneous issuance of an ultimatum and a call for an Extraordinary Meeting. Such a meeting almost invariably carried an implicit (and deliberately unconfirmed) threat of unilateral legislation or action. At the last moment, by accepting OPEC's demands "in principle," the oil companies conceded defeat. OPEC postponed unilateral action and continued negotiations, which usually culminated in an agreement incorporating a substantial number of OPEC's demands.

The Extraordinary Meeting was rescheduled because of in-group squabblings between the hardliners, Kuwait, Iraq, and Libya, whose oil ministers constituted OPEC's negotiating team, and the moderates, Saudi Arabia, Abu Dhabi, and Iran. These two groups were fundamentally at variance on the issue of price adjustments. The moderates wanted only to modify Geneva I, while the hardliners, spearheaded by Libya, were ostensibly interested in destroying it.

Iran was in the midst of delicate negotiations with its consortium and was not interested in jeopardizing them by confronting the oil companies over what it considered to be "minor issues" of price adjustment. Saudi Arabia and Abu Dhabi were afraid that a confrontation with the oil industry over price adjustment might impede the buy-back negotiations whose successful conclusion was a vital precondition for implementing the participation agreements (which will be discussed shortly).

OPEC took a decisive step to resolve its in-group conflict. Two representatives from moderate states, Zaki Yamani of Saudi Arabia and Nigerian Oil Minister Shettima Ali Monguno, were added to the negotiating team. The OPEC team then met with its counterparts from the oil industry in Geneva on May 28, and an agreement was reached on June 1, 1973. The Geneva II agreement, which revised Geneva I, involved Iran, Saudi Arabia, Iraq, Qatar, Kuwait, and Abu Dhabi; the Libyan agreement of 1972; and the Nigerian agreement of June 1972. The other three OPEC members, Venezuela, Indonesia, and Algeria, had their own systems of fixing prices. The main provisions of the Geneva II agreement are as follows:[13]

(1) Basically it preserved the framework of Geneva I, a clear victory for the oil industry and a defeat for the hardline oil states, which wanted to do away with Geneva I.

(2) The oil companies yielded to the OPEC contention that the 5.8 percent increase to compensate the dollar devaluation negotiated under Geneva I was not enough since the oil states' loss of purchasing power amounted to 11.7 percent. The oil industry agreed to provide Libya and Nigeria (the Mediterranean states) an upward adjustment of 11.9 percent for posted prices that were effective prior to the February 1973 dollar devaluation and about 5.7 percent on prices posted by the companies on April 1, action in line with the original Geneva formula. For the Gulf states, the price adjustments were to take the course shown in Table 6. In light of the January increase shown in this table, the price of Libyan crude oil was to increase by 47.3¢/b, or 12.5 percent, to $4.250/b, and the government take by 29¢/b, to roughly $2.510/b.

(3) In an attempt to measure currency fluctuations, Geneva II expanded the existing group of nine reference currencies to eleven. The two additions were the Canadian dollar and the Australian dollar.

(4) The new formula was more sensitive and responsive to changing international monetary relationships.

(5) Prices were to be reviewed on the twenty-third of each month instead of quarterly as under Geneva I. The parties agreed to adjust prices in case of an upward or downward fluctuation of 1 percent, in contrast to the agreement under Geneva I to adjust prices in case of 2 percent fluctuations. It was also agreed that the prices would not be allowed to fall below the minimum amount agreed to under the Teheran agreement of 1971.

Table 6. Increase in Posted Price, Government Take, and Tax-Paid Cost for Gulf States at the Conclusion of Geneva II

	Jan. 1	April 1	June 1	Increase, Jan.-June		Increase, April-June	
	($/b)	($/b)	($/b)	¢/b	%	¢/b	%
Posted price	2.591	2.742	2.898	30.7	11.85	15.6	5.7
Government take	1.516	1.607	1.700	18.4	12.1	9.3	6.1
Tax-paid cost	1.616	1.707	1.800	18.4	11.4	9.3	5.4

Source: Supplement to *Middle East Economic Survey,* June 1, 1973, p. 2.

The Geneva II formula was expected to remain in effect until 1975, when the Teheran and Tripoli agreements were to expire. But under the chaotic conditions following the Arab oil embargo and the unilateral price increases of 1973, OPEC eventually abandoned the Geneva II agreement.

In both negotiations that followed the dollar devaluations, the impact of the growing sellers' market continued to be felt in the gradual erosion of the oil industry's bargaining position. The negotiations culminating in the Geneva I agreement were characterized by the routinization of regional cooperation among the oil states. The final outcome of Geneva I reflected the aims of the moderates who shaped it. The hardline states, such as Libya, had to take a back seat. But the hardliners' decision to adjust the value of their currency unilaterally, along with the Venezuelan revaluation of the bolivar, markedly strengthened the overall bargaining position of the oil states. Even though the Geneva II negotiations were characterized by conflict between the moderate and hardline oil states, such conflict remained manageable and OPEC members were able to extract concessions from the oil industry on almost all of their demands. The burden of this success, however, rested with the uncertainities of the sellers' market.

Through Geneva I and II, OPEC members were successful in considerably broadening the scope and dimension of their concern about international inflation and its erosive impact on their buying power and on terms of trade between their economies and those of the Western industrial countries. It also became abundantly clear that in view of their growing desire to seek long-term solutions to these problems, the oil states would not be content with just negotiating periodic price increases with the oil industry and would be forced to establish in the near future avenues to deal directly with the Western industrial countries.

The Geneva I and II agreements also marked routinization of leapfrogging in the prices of crude. OPEC members continued to introduce periodic increases in the prices of their commodity as long as international inflation remained unchecked. Leapfrogging, however, continued to boomerang, as increased inflation led in turn to further erosion of the buying power of the oil states, who responded by further leapfrogging. This vicious cycle had no apparent end in sight and its effects were intensified in the years that followed.

The Whirlwinds of the Sellers' Market

Under this rubric, certain radical and unprecedented changes will be examined that were made possible by the growing uncertainties of the sellers' market. These changes were radical in the sense that they clearly marked the emergence of the oil states as the advantaged actors and the oil industry as the disadvantaged actors. Such changes were unprecedented because they enhanced the importance of the oil states not only in influencing the economic well-being of the industrial countries but also in sustaining the strategic status quo between the East and the West. One unprecedented aspect of these changes was OPEC's decision to remove the oil industry once and for all, from future negotiating sessions that would determine the timing and magnitude of price escalations. The rationale of these periodic escalations continued to be international inflation. But the removal of the oil companies, which served as an effective check on OPEC's desire to introduce unrealistic increases in the prices of crude, inevitably meant that these increases would be based only on the oil states' desire to minimize the impact of international inflation on their purchasing power, not on the implications of these escalations on the world economy.

Unilateral Price Increases The ink on the Geneva II agreement was barely dry when OPEC signaled its intentions to boost the prices of crude again. The oil states were becoming increasingly disgruntled with the continued instability of world currencies and the increasing rate of inflation. They had negotiated the Teheran agreement on the assumption that the U.S. dollar would remain stable and that the rate of inflation in the industrial countries, which had maintained a modest pace during the 1960s, would not exceed 2.5 percent. Both these assumptions, turned out to be incorrect by 1973. Between 1971 and 1973, the U.S. dollar was devalued twice, and in the early 1970s inflation began to take off, with export prices of OECD manufactured goods rising 6 to 8 percent annually. As indicated in Figure 3, the demand for oil also continued to rise, the most convincing evidence needed by the oil-exporting countries that further price increases were needed. They felt that the higher prices charged consumers by the oil industry were not reflected in proportionally higher returns in their oil revenues. They also were of the view that the profit-sharing ratio on realized prices between governments and the oil industry, which at the time of the Teheran agreement remained at roughly 80-20 in the governments' favor, should be reduced to 64-36.[14] OPEC members also believed that the Teheran agree-

Figure 3. World Oil Consumption 1960-1978

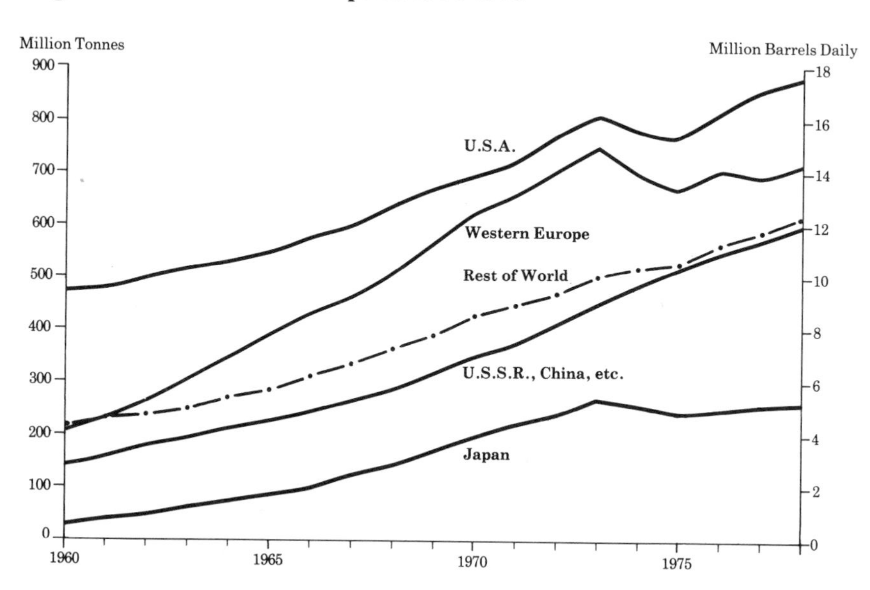

Source: *BP Statistical Review of the World Oil Industry, 1978.*

ment, since it was no longer compatible with prevailing market conditions or with galloping inflation, was in need of extensive revision. In an interview for the *Middle East Economic Survey,* Yamani outlined OPEC's thinking on amending the Teheran agreement along these lines. He stated that if posted prices were to resume their function as realistic tax reference prices, they should be substantially increased, that a mechanism should be devised that would maintain posted prices above realized prices at all times, and that the 2.5 percent annual increase for inflation and the 5¢/b negotiated in the Teheran agreement should be amended to reflect more realistically the prevailing rates of worldwide price inflation.[15] It was apparent that OPEC was intent on restoring the profit take of the oil companies, in percentage as well as in absolute cents per barrel, to the levels prevailing at the time of the Teheran agreement so that the resultant profit could revert to the governments. Prevailing market conditions made OPEC members abundantly confident that, in the absence of a negotiated settlement with the oil industry, the oil states could take charge and set the prices themselves.

During the two-day Special Ministerial Conference in Vienna, September 15-16, 1973, OPEC issued an invitation to the oil industry for a meeting on October 8 to renegotiate the basic oil price agreement. A noteworthy aspect of this "invitation" was not missed by the *Oil and Gas Journal,* an authoritative partisan of the oil industry, when it observed that the issuance of "the dictatorial summons" and the setting of the October date without consulting industry representatives were indeed demonstrations of new "OPEC power."[16]

On October 6, 1973, hostilities erupted between Arab and Israeli forces along the Suez Canal and the Golan Heights. Against this backdrop, negotiators of the Persian Gulf states and the oil industry began their parlays on October 8. The industry's initial offer was comprised of a 45¢/b increase in posted prices and the substitution of a new inflation adjustment index to replace the 2.5 percent inflation factor that had been part of the Teheran agreement. The oil companies also offered to introduce a sulfur premium which was to be applied against the higher-quality crudes in the Gulf. OPEC's demands, however, included a $3/b increase in posted prices, and even this increase was subject to further escalation on the basis of an inflation index. The Gulf states also sought an agreement whereby the posted prices would remain at 40 percent above market prices. Because of the magnitude of these demands, the negotiating team from the oil companies asked for a two-week recess to consult with the major consumer governments.[17]

A second negotiating session was never held. On October 16, the Persian Gulf members of OPEC announced a unilateral increase in posted prices which raised the price of Arabian Light, the key crude in the Gulf area, from $3.11 to $5.119/b, an increase of 170 percent. The oil states also declared that, in future, posted prices were to be determined by actual market prices in the Gulf as well as in other areas and that the relationship between the two prices was to be the same as the one that had prevailed before the Teheran agreement. In addition, upward and downward corrections in posted prices were to be brought about when the actual market prices of crude oil varied by more than 1 percent from the announced prices.[18] The Gulf states also declared that if the oil industry balked at accepting the new prices, they would sell their commodity on the open market based on the new market price of Arabian Light.

The October 16 announcement established OPEC as the sole determiner of oil prices and ended, once and for all, the long-established tradition of consultation and negotiation between the oil states and the industry prior to any upward move in prices of crude. The timing of this announcement could not have been better for the purposes of OPEC, for the oil industry was in no position to defy their waxing power. This announcement effectively transformed the oil states into independent merchants. Under normal conditions of supply and demand this transformation would have favored the buyers of crude since the oil states had to compete for buyers. But October 1973 was a far from normal time, with war raging in the Middle East and with markets thirsty for high-quality, low-sulfur, and short-haul crude, much of which was then denied consumers by the war. The October 16 announcement was revolutionary also from another point of view. It introduced such high increases in crude oil prices

Table 7. Crude Oil Price Increases, October 1-31, 1973 ($/barrel)

Source/Type	Gravity	Posted or Tax-reference Price FOB Loading Port Oct. 1	Oct 31.
Persian Gulf			
Abu Dhabi			
Murban	39	3.144	6.045
Umm Shaif	37	3.110	5.538
Zakum	40	3.185	5.964
Iran			
Basrah	35	2.977	5.061
Light	34	2.995	5.091
Heavy	31	2.936	4.991
Iraq			
Kuwait	35	2.884	4.903
Qatar			
Dukhan	40	3.143	5.343
Marine	36	3.037	5.163
Saudi Arabia			
Light	34	3.011	5.119
Medium	31	2.884	4.903
Heavy	27	2.725	4.632
Mediteranean			
Libya	40	4.604	8.925
Algeria	43	5.000	5.000[a]
Arabian Light (Sidon)	34	4.205	7.148[b]
Kirkuk	36	4.243	7.213[b]
South America			
Ecuador	28	3.600	5.250
Venezuela	11-39	4.610Av.	4.610[c]
Nigeria			
Light	34	4.287	4.291
Blend	27	4.148	4.148
Indonesia			
Kasim	43.5	5.000	5.000
Minas	34	4.750	4.750

Source: *Oil and Gas Journal*, Oct. 29,1973, p. 50.

[a]Spot $7.00/b.

[b]Estimated.

[c]Adjustment due Nov. 1 to take account of OPEC increases.

that the oil companies, in order to sustain their accustomed profit margin, had to pass it on promptly to the consumers. Consequently, consumers in both industrial and nonindustrial countries bore the brunt of OPEC's increases. Table 7 provides a picture of price increases from October 1 through October 31, 1973.

On October 17, only one day after the fateful unilateral price declaration by OPEC, the ten members of the Organization of Arab Petroleum Exporting Countries (OAPEC) imposed an oil embargo on their crude supplies to the United States and the Netherlands. These two countries were to be deprived of Arab oil as a punishment for what the Arabs perceived as their unequivocal support of Israel. Supplies to other Western industrial countries were also to be reduced. By imposing the oil embargo, the Arab oil states were using their commodity to try to induce the United States and its Western allies to apply pressure on Israel to resolve the obdurate Arab/Israeli conflict. A detailed examination of the oil embargo will be made in the next chapter. Initially, the participating Arab oil states decided to cut their production by 5 percent or more for October; this initial cut was to be followed by successive cuts of an additional 5 percent each month until Israel withdrew from territories occupied in the war of 1967.

Against this background, OPEC held meetings in Vienna, November 17-19, with representatives of the oil companies: Exxon, Royal Dutch-Shell, Standard Oil, Atlantic Richfield, and British Petroleum. The participants' obsession with their respective objectives was so overwhelming that one authoritative source characterized these meetings as almost "a dialogue of the deaf." OPEC members were primarily interested in discussing future mechanisms for determining market price levels for crude oil on which they could base posted prices in accordance with their October 16 pronouncement. The representatives of the oil companies, on the other hand, wanted to express their disdain of the whole system of maintaining posted prices at a constant 40 percent above market prices. They claimed this system had a potential for leapfrogging escalations and a tendency to freeze the profits of the oil companies on crude sales at a constant level. The oil men pleaded with OPEC to develop "gradual and predictable" mechanisms of price escalation in the interest of the world economy. They also urged the oil states to allow a relatively long time between posted price revisions. OPEC members saw the position of the oil representatives as a covert attempt to bring back a long-term Teheran-type agreement which would minimize the upward movement in posted prices, and they wanted no part of it. Evidently OPEC members were enjoying their newly-acquired ability to make unilateral price adjustments and were totally disinterested in establishing any long-term pricing mechanisms which, under the mounting uncertainties of the sellers' market, were bound to be outmoded in a far shorter period of time than ever before.

The delegates from OPEC countries were divided between advocates of two widely diverging pricing strategies. The price hawks favored "playing the market for all it was worth," while the moderates argued for "taking it easy for the time being." Concerning a long-range pricing mechanism, however, there was a consensus that future determination of crude oil prices should be market-oriented. On the short-range question of determining the market price on which posted prices were to be based at any given time, the members decided to base it on government f.o.b. sales to third parties of certain marker crudes in the Gulf: namely, Arabian Light, Iranian Light, and Iraq Basrah. In the long run, however, the oil states wanted to develop a global pricing structure that would include posted prices throughout the OPEC area, and a comprehensive mechanism and procedure for price changes that would insure internal harmony between crudes of various gravities and locations. The conferees also decided that the prices should be reviewed on a quarterly basis.[19]

The November 17 meeting was the last time the oil companies and the oil states met with the ostensible purpose of discussing crude prices. The uncertainties of the sellers' market took their toll in the demise of the once-powerful oil companies as the negotiating partners of OPEC on oil pricing. From that time on, OPEC was decisively the advantaged actor while the oil companies were reduced to the role of technician-managers with regard to upstream operations. In the downstream operations of the oil business, however, the oil companies retained their unquestioned authority. Little, if any, breakthrough was made by the state-owned companies of the oil states. For this reason, the issue of oil industry profits remained largely unresolved. How much profit should the oil companies be allowed? Was there any way OPEC could set a "legitimate" profit margin for the oil industry? OPEC was to grapple with these questions through the year 1974.

The demise of the oil companies as protagonists meant that the oil states were left with the prerogative of unilaterally adjusting the prices of their commodity at will. But this prerogative produced a negative externality whose effects were to intensify whenever the oil states introduced large price increases. The unilateral price increases also facilitated the development of a tradition of direct dealings between the petroleum-exporting countries and the industrial consuming nations. A closer look at these two phenomena is in order.

The Effects of Unilateral Price Increases OPEC's unilateral price increase of October 16 was the beginning of a tradition whereby the oil states were to introduce unrealistic increases in oil prices under tight market conditions, disregarding the implications of their action for the world economy. The consuming nations in general suffered considerably. The price increases also created a "catch-22" situation for the oil states whereby, instead of insulating themselves from the effects of international inflation, they increased world inflationary rates, which further eroded their purchasing power.

At the end of their meeting in Teheran on December 23, 1973, the Gulf members of OPEC again raised the posted prices of Arabian Light 34° marker crude from $5.119 to $11.651/b, effective January 1, 1974. This was an increase of 227 percent, making the cumulative increase from October through December more than 384 percent. Tables 8 and 9 provide an overview of escalations in posted prices and in government take for the six Gulf states from 1971 through January 1974.

The December conference was concerned with three major issues. First and foremost, the oil states were still preoccupied with their quest for a long-term pricing mechanism. They were undecided as to how to determine market prices: on the basis of a netback formula from refined product sales in consumer markets, on the basis of f.o.b. realizations from their own oil transactions, or on the basis of a comparison with the costs of alternative sources of energy. The second issue was the excessive profits made by the oil industry in a volatile market. In October, by deciding to keep posted prices at 40 percent above market prices, OPEC attempted to maintain the profit-sharing ratio between companies and governments at 80-20 in favor of the oil states. In a rapidly changing market, however, this system created a spiral and required immediate correction. The third obdurate issue faced by OPEC members was international inflation.

No long-range decision was made on the issue of a pricing mechanism. But this conference was marked by the beginning of the Saudi Arabian role as a price moderate. Iran was a leading price hawk and during the initial bargaining session pressed for a price considerably higher than the figure eventually agreed upon. Yamani, the Saudi oil minister, fearing the devastating effects the increased price would have on the free world's economy, refused to go along with such an increase. The $7/b increase in government take (see the January 1, 1974, figure in Table 4.3) was a compromise between the Iranian and Saudi positions. At a grandstand press conference, the Shah of Iran announced that the era of extraordinary progress and income based on cheap oil had ended. "What we actually want," he said, "is not to be cheated over the price of a commodity which in any case will be used up, maybe, in thirty years, and secondly, to gain new sources of energy which will replace this precious oil everywhere as much as possible."[20] The Shah observed that the $7/b government take decided upon at the Teheran meeting was based on initial estimates of the comparative costs of alternative forms of energy, such as oil from shale or liquefaction or gasification of coal. He proposed that OPEC and the Organization for Economic Cooperation and Development (OECD) initiate joint studies to ascertain the real costs of extracting oil from alternative sources, the results of which would serve as a comparative yardstick for conventional oil prices.

Concerning the excessive profits of the oil industry, the oil states were leaning in favor of adjusting tax rates to eliminate what they considered excess

Table 8. Evolution of Posted Prices for Main Gulf Crudes, 1971-1974 ($/barrel)

	1971			1972	1973									1974
Source/Type	Pre-Feb. 15	Feb. 15	June 1	Jan. 20	Jan. 1	Apr. 1	June 1	July 1	Aug. 1	Oct. 1	Oct. 16	Nov. 1	Dec. 1	Jan. 1
Saudi Arabia														
Light	1.800	2.180	2.285	2.479	2.591	2.742	2.898	2.955	3.066	3.011	5.119	5.176	5.036	11.651
Medium	1.680	2.085	2.187	2.373	2.482	2.626	2.776	2.830	2.936	2.884	4.903	4.957	4.822	11.561
Heavy	1.560	1.960	2.064	2.239	2.345	2.481	2.623	2.674	2.755	2.725	4.633	4.684	4.557	11.441
Iran														
Light	1.790	2.170	2.274	2.467	2.579	2.729	2.884	2.940	3.050	2.995	5.341	5.401	5.254	11.875
Heavy	1.720	2.125	2.228	2.417	2.527	2.674	2.826	2.881	2.989	2.936	4.991	5.046	5.006	11.635
Kuwait														
Kuwait	1.680	2.085	2.187	2.373	2.482	2.626	2.776	2.830	2.936	2.884	4.903	4.957	4.822	11.545
Abu Dhabi														
Murban	1.880	2.235	2.341	2.540	2.654	2.808	2.968	3.026	3.140	3.084	6.045	6.113	5.944	12.636
Marine (Umm Shaif)	1.860	2.225	2.331	2.529	2.642	2.796	2.955	3.013	3.126	3.070	5.537	5.599	5.446	12.086
Zakum	—	—	—	—	—	—	—	—	—	—	5.964	6.031	5.865	12.566
Iraq														
Basrah	1.720	2.155	2.259	2.451	2.562	2.711	2.865	2.921	3.031	2.977	5.061	5.117	4.978	11.672
Qatar														
Dukhan	1.930	2.280	2.387	2.590	2.705	2.862	3.025	3.084	3.200	3.143	5.834	5.899	5.737	12.414
Marine	1.830	2.200	2.305	2.501	2.614	2.766	2.923	2.980	3.092	3.037	5.503	5.563	5.412	12.013

Source: *Middle East Economic Survey,* Dec. 28,1973, p.5.

Table 9. Evolution of Government Take for Main Gulf Crudes, 1971-1974 ($/barrel)

	1971			1972	1973									1974
Source/Type	Pre-Feb. 15	Feb. 15	June 1	Jan. 20	Jan. 1	Apr. 1	June 1	July 1	Aug. 1	Oct. 1	Oct. 16	Nov. 1	Dec. 1	Jan. 1
Saudi Arabia														
Light	0.989	1.261	1.325	1.448	1.516	1.607	1.702	1.736	1.804	1.770	3.048	3.083	2.998	7.008
Medium	0.930	1.203	1.265	1.384	1.450	1.537	1.628	1.661	1.725	1.694	2.917	2.950	2.868	6.954
Heavy	0.843	1.106	1.169	1.302	1.367	1.449	1.535	1.566	1.627	1.597	2.754	2.785	2.708	6.881
Iran														
Light	0.983	1.250	1.313	1.430	1.497	1.588	1.683	1.717	1.783	1.750	3.172	3.208	3.119	7.133
Heavy	0.944	1.222	1.285	1.400	1.466	1.555	1.647	1.681	1.746	1.714	2.960	2.993	2.969	6.988
Kuwait														
Kuwait	0.985	1.231	1.293	1.406	1.472	1.559	1.650	1.683	1.747	1.716	2.939	2.972	2.890	6.966
Abu Dhabi														
Murban	1.005	1.272	1.337	1.458	1.527	1.620	1.717	1.752	1.821	1.787	3.582	3.623	3.521	7.578
Marine (Umm Shaif)	0.966	1.239	1.288	1.391	1.460	1.553	1.650	1.685	1.753	1.720	3.192[a]	3.229	3.137	7.162
Zakum	—	—	—	—	—	—	—	—	—	—	3.451[a]	3.492	3.391	7.453
Iraq														
Basrah	0.933	1.240	1.303	1.419	1.487	1.578	1.671	1.705	1.772	1.739	3.002	3.036	2.952	7.010
Qatar														
Dukhan	1.052	1.316	1.381	1.493	1.546	1.641	1.740	1.776	1.847	1.812	3.443	3.482	3.384	7.432
Marine	0.924	1.196	1.260	1.351	1.464	1.556	1.651	1.686	1.754	1.720	3.215	3.252	3.160	7.162

Source: *Middle East Economic Survey,* Dec. 28, 1973, p. 6

Government take comprised 12.5% expensed royalty plus 55% income tax calculated on the full posted price (except for appropriate OPEC discount for the pre-Feb. 1971 column). The following estimates for production costs per barrel were used: Arabian Light and Medium: 11¢ in 1971, 10¢ in 1972-74; Arabian Heavy: 15¢ in 1971, 10¢ in 1972-74; Iranian Light and Heavy: 12¢; Kuwait: 6¢; Abu Dhabi Murban: 15¢; Abu Dhabi Marine (Umm Shaif): 20¢ in 1971, 30¢ in 1972-74; Abu Dhabi Zakum: 30¢; Iraq Basrah: 12¢; Qatar Dukhan: 12¢ in 1971, 14¢ in 1972, 17¢ in 1973-74; Qatar Marine: 25¢ in 1971, 30¢ in 1972, 22¢ in 1973-74.

[a]Prior to Oct. 16, 1973, government take in Abu Dhabi Marine (Umm Shaif) was averaged out with Zakum, which had no separate posting at that time.

company profits and of calculating taxes and royalties on realized market prices rather than on posted prices. On the issue of inflation, too, no long-term decision was made. The Shah suggested that OPEC and the OECD countries deliberate over two options: either the purchasing power of oil states should keep pace with the rates of inflation in the industrial countries or a system must be evolved whereby the oil states would sell oil at a fixed price provided they could purchase goods from industrial countries at a fixed price. This linkage between inflation and the purchasing power of the oil states, which had remained at the core of price increases since the Teheran agreement of 1971, became an institutionalized pattern for future price increases introduced by OPEC throughout the 1970s. Moreover, these exorbitant increases turned out to be as much of a nuisance for OPEC as was Midas's ability to turn every object into gold by his mere touch: a source of riches turned into curses or no-win situations. Similarly, the ease with which the oil states could raise oil prices to safeguard their buying power from inflation turned out to be one of the leading causes of further inflation. During the 1970s, OPEC remained preoccupied with the means to use its Midas touch wisely.

In the 1960s, when OPEC's chief obsession was to raise the prices of crudes, the issue of how much increase never really became germane primarily because of the inability of the organization to raise prices. In the 1970s, when OPEC succeeded in loosening the oil industry's grip on the oil market and in setting the prices of crudes unilaterally, the issue of how much increase was bound to become a crucial one. The importance of this issue rested on the fact that the non-oil-producing developing countries were being crushed under the growing burden of oil-related debts. For example, according to one estimate, during 1974 India was expected to pay some $1.240 billion for imported crude oil needs, which was about 40 percent of the country's potential export earnings and twice as much as its prevailing foreign exchange reserves.[21] Even the ability of the Western industrial countries to pay the swiftly rising prices of oil was not infinite. Based on pre-embargo production rates, according to another source, OPEC members' income, which stayed at $30 billion on the basis of posted price levels of October 1, increased to $51 billion as a result of the price increases of October 16 and was expected to far exceed $100 billion in 1974.[22] According to James Grant, the cif cost of oil imports from OPEC countries was $20 billion in 1972, $36 billion in 1973, and $100 billion in 1974. OPEC's current account surpluses for the same years rose from $1.6 billion to $6.1 billion and $66 billion. At the same time, the cost of oil imports by developing countries rose from $3.7 billion in 1972 to $5.2 billion in 1973 and $15 billion in 1974.[23] Obviously, some means had to be found to provide respite for the world economy from seemingly unending rounds of price escalations.

Before the unilateral price increase of October 16, the oil companies served as moderators of OPEC's pricing behavior. But the removal of the oil companies from exchanges did not eliminate the vital need for moderation on the issue of price increases. Starting in 1974, Saudi Arabia emerged as the leading force for price moderation within OPEC. The Saudi decision to adopt this role stemmed from the fact that that country had the largest quantities of proved oil reserves, and in this capacity its economic prosperity was inextricably linked to the economic stability and healthy growth of the Western industrial economies. A related and equally important factor was that the accumulation of petrodollars in the hands of the oil states, especially the Arab oil states, was causing enormous consternation in the Western industrial world. The predominant feeling in the West was that the oil states could not be trusted to behave responsibly in the international money market, an opinion based on their continued disregard of the impact of their unrealistic and intermittent price increases on the world economy. If the oil states remained as oblivious to the implications of their economic actions in the international financial market as they had been about raising oil prices, the Western industrial states believed the non-Communist economies would face an even more uncertain future. Such feelings were bound to complicate recycling[24] of petrodollars, which were viewed as crucial not only to the survival of the Third World economies but also to the potential growth of the industrial economies. Under these circumstances, Saudi Arabia, the possessor of the largest share of petrodollars, was in a mood to be cautious with the world economy by making it very difficult for its OPEC allies to continue introducing unrealistic price increases.[25]

Beginning in 1974, OPEC meetings were marked by in-group squabblings between the price hawks, an informal and ad hoc coalition of Iran, Libya, and Algeria, which were frequently joined by Iraq and Nigeria, and the price moderates, led by Saudi Arabia and regularly joined by the UAE, Kuwait, and Qatar. As the economic slump caused by the recession of 1974-1975 became acute in the non-Communist economies, OPEC members found it increasingly difficult to raise oil prices not only because of a downturn in oil demand but also because of the mounting Saudi opposition to price increases before the effects of recession disappeared from the industrial economies.

The effects of the 1973 quantum price increase were not yet apparent at the beginning of 1974. The U.S. dollar remained strong. In light of the Geneva I and II currency agreements, under which postings in the Gulf, Libya, and Nigeria were supposed to move up or down on a monthly basis in line with an index of the fluctuation of the U.S. dollar against currencies of eleven industrial countries, a reduction of 6 percent in posted prices was required. At the January 7-9 OPEC conference, this reduction in posted prices became an issue of contention between the price moderates and the hawks. The moderates, particularly Saudi Arabia and Kuwait, favored such a reduction, while the hawks vehemently opposed it. Libya argued that gains from the appreciation of the dollar were being more than offset by inflation. As an expression of their continuing search for a long-range pricing mechanism, the oil states instructed the OPEC Economic Commission to study urgently the issue of posted prices with a view to establishing a pricing system for crude oil in the long run. The Commission's study was to focus on competitive price levels of alternate energy sources; on comparative advantages of petroleum over other fuels, such as ease of handling and petrochemical derivatives; on the effects of oil prices on world balances of payment; and especially on protecting the purchasing power of the oil states.

In March, the price hawks based their advocacy of further price hikes on recommendations made by the Economic Commission. According to these recommendations, market conditions justified an increase of $2.67/b. Such an increase would raise the price of 34° Arabian Light from $11.651 to $14.30/b and would add between 85¢ and $1.60/b to the government take. The Commission's recommendations were favorably received by all member states save Saudi Arabia. In fact, Kuwait and the UAE, which usually followed the Saudi policy of moderation on the price issue, were critical of Saudi public advocacy of the price freeze. Kuwait and the UAE were of the view that the Saudi stance signaled customers that OPEC would not be able to raise prices and caused them to submit low bids at auction sales in those two countries. Saudi Arabia, on the other hand, maintained that the escalations in posted prices implemented on January 1 were inordinate, were possible only because of exigencies created by the oil embargo, and were a threat to the world economy. The Saudis continued in their determination to disallow further price increases at the OPEC conference in Quito, Ecuador, in June. After an acrimonious debate, OPEC announced an extension of the price freeze. Saudi Arabia succeeded in extracting this extension only by threatening to reduce prices unilaterally and to increase production if other oil states implemented the proposed increase. The conferees agreed to raise the royalty rates from 12.5 percent to 14.5 percent and to increase the government take on Iranian Light from $7.133 to $7.240/b. These increases were to be effective July 1.[26] In the sellers' market, and with the removal of the oil industry from the price-fixing process, the oil states were feeling the burden of responsibility to the world economy, the survival of which was intrinsically linked with their own. Not all states were expected to alter their behavior by trying to balance their own economic interests with those of the world community. But the fact that some oil states were increasingly forced to play this balancing game was a sign that moderation was likely to play a meaningful role in OPEC's decision in the days to come.

During the third quarter of 1974, the world oil market was beginning to show signs of growing surpluses. The impact of this phenomenon was felt in July, when Iran offered to sell 215,000 b/d for at least 93 percent of the posted price. At the asking price, only 8,000 b/d found a buyer. The government of Kuwait encountered a similar problem when it tried to auction 1.25 million b/d for the last half of 1974 and all of 1975 at minimum bids of 95.7 percent of the posted price. The Mediterranean oil producers were also forced to shave their prices. OPEC's mounting concern over market conditions was reflected in two noteworthy recommendations agreed upon at a technical meeting of the national oil companies (NOCs) of OPEC. These recommendations were to be submitted to the governments concerned for appropriate action at the regular meeting of OPEC. The OPEC technocrats agreed not to sell their crude oil below 93 percent of the posted price for 34° Arabian Light and to make appropriate adjustments for freight differential and quality factors such as gravity and sulfur content. The second recommendation was aimed at reducing the growing oil surplus, which was estimated to be between 2 and 4 million b/d. The technocrats consented to a two-pronged attack on the problem of surpluses. First, they felt that some measures aimed at controlling production were warranted. The second corrective measure, which was more realistic in that it reflected the prevailing thinking of decision makers at the highest echelon of OPEC, needs elaboration. The integrated oil companies (mostly international majors) with access to "participation crude," which was available to them under the relatively favorable buy-back arrangements worked out in 1973, were underselling those companies that were compelled to buy from state-owned oil companies, such as Sonatrach of Algeria, Iraq National Oil Company (INOC), or Petramina of Indonesia. This situation created a two-tier price system. The OPEC technocrats' corrective measure was aimed at unifying the two-tier price system by bringing the posted prices of participation crude to the international majors more in line with market prices. But any resolution of problems created by participation crude was contingent upon the successful conclusion of the participation negotiations between Aramco and the government of Saudi Arabia. If these negotiations produced an agreement resulting in a 100 percent takeover of Aramco by the Saudis, requiring the Aramco partners to buy crude at full market prices and to receive cash payment for their services, then it would also eliminate the participation crude altogether.

By mid-1974, signs of a new crisis for OPEC were beginning to emerge. The world oil market was showing the real effects of OPEC's January 1 increases. The industrialized countries were forced to adopt conservation policies which brought about a noticable reduction in their levels of consumption. There were reports of a surplus of about 2-4 million b/d in immediately usable producing capacity. A U.S. study presented in June at a meeting of the Energy Coordination Group, a coordinating body of twelve industrialized nations formed during the Washington Energy Conference of February 1974, predicted that oil supply was likely to exceed demand in the next twelve months. After noting that the prevailing combination of free world production and price levels was unsustainable, the study stated that reduction in output, price, or both seemed likely during the coming summer. The study indicated that in April of 1974 the unused productive capacity of OPEC was 4.5 million b/d, of which 4

Table 10. Crude Oil Production in OPEC Member Countries, 1970-1977 (1,000 b/d)

Producer	1970	1971	1972	1973	1974	1975	1976	1977
Algeria	1,029.1	785.4	1,062.3	1,097.3	1,008.6	1,020.3	1,075.1	1,152.3
Ecuador	4.1	3.7	78.1	208.8	177.0	160.9	187.8	183.4
Gabon	108.8	114.6	125.2	150.2	201.5	223.0	222.8	222.0
Indonesia	853.6	892.1	1,080.8	1,338.5	1,374.5	1,306.5	1,503.6	1,686.1
Iran	3,829.0	4,539.5	5,023.1	5,860.9	6,021.6	5,350.1	5,882.9	5,662.8
Iraq	1,548.6	1,694.1	1,465.5	2,018.1	1,970.6	2,261.7	2,415.4	2,493.0
Kuwait	2,989.6	3,196.7	3,283.0	3,020.4	2,546.1	2,084.2	2,145.4	1,969.0
Libya	3,318.0	2,760.8	2,239.4	2,174.9	1,521.3	1,479.8	1,932.6	2,063.4
Nigeria	1,083.1	1,531.2	1,815.7	2,054.3	2,255.0	1,783.2	2,066.8	2,085.1
Qatar	362.4	430.7	482.4	570.3	518.4	437.6	497.3	444.6
Saudi Arabia	3,799.1	4,768.9	6,016.3	7,596.2	8,479.7	7,075.4	8,577.2	9,200.0
UAE	779.6	1,059.5	1,202.7	1,532.6	1,678.6	1,663.8	1,936.4	1,998.7
Venezuela	3,708.0	3,549.1	3,219.9	3,366.0	2,976.3	2,346.2	2,294.4	2,237.9
Total	23,413.0	25,326.3	27,094.4	30,988.5	30,729.2	27,192.7	30,737.7	31,398.3

Source: *Annual Statistical Bulletin*, 1975 and 1976 (Vienna: OPEC).

million b/d were produced by the Arab countries, particularly Kuwait and Libya. This study estimated that while the non-Communist world's energy needs were expected to revolve around 31.4 million b/d in the second half of 1974 and 32.1 million b/d during the first half of 1975, OPEC production for the same period was expected to grow to 32.9 million and 34.8 million b/d, respectively.[27] OPEC did not welcome this news, since excessive production was bound to depress the prices of crude oil.

Some oil states responded to this growing uncertainty in the oil market by instituting crude production cutbacks to forestall lower prices. Kuwait was in the vanguard of such a policy. Kuwait's lead was followed by an announcement from the government of Venezuela of a crude reduction of about 100,000 b/d. Moreover, Libya was reported to have quietly imposed conservation-oriented reductions on several companies. Table 10 provides an overview of production cutbacks for 1973-1974. It should be noted that Saudi Arabia and Iran raised their production, though not significantly, while Libya, Kuwait, and Iraq lowered their levels of output.

The excess supply situation exacerbated existing tensions within the rank and file of OPEC, as witnessed at the Quito meeting. It was apparent that in order to sustain price levels, the oil states would be forced to develop an elaborate system of production programing. It will be recalled that OPEC had failed to develop such a mechanism once before, during the 1960s, under the buyers' market. In 1974, even though the oil market remained a sellers' market, OPEC was not expected to adopt production programing yet for two reasons. First, there were still no institutional means to police such a program, and second, Saudi Arabia, as the country with the largest oil reserves, continued to oppose production programing. And even as dwindling demands were threatening the continuation of high prices, the chances of Saudi willingness to go along with production programing were at best slim. In fact, extrapolating from the continued opposition to the January 1 level of prices, that country was expected to defeat any efforts to institutionalize production programing, since the absence of such an arrangement promised a lowering of prices, an outcome desired by the Saudis in the past. The excess supply situation was also expected to hamper the buy-back negotiations, without whose successful conclusion the implementation of participation agreements was not possible. Even if OPEC were to incorporate production programing, in the two-tier price system which enabled the international majors to undersell the NOCs, the oil states were still expected to encounter considerable difficulty finding customers at the higher tier of market prices.

As expected, OPEC members failed to reach an agreement on production programing during their two-day Extraordinary Meeting. Since, however, a number of oil states had already introduced voluntary production cutbacks in order to stabilize the prices of their crudes, no immediate decision on this issue was deemed necessary. The oil states agreed to extend their price freeze to the fourth quarter of 1974, thus keeping the posted price of Arabian Light marker crude at $11.651/b. On participation crude, which in most cases amounted to 40 percent of production, OPEC once again increased royalty rates, from 14.5 to 16.67 percent, and tax rates, from 55 to 65.75 percent. This decision raised government take on participation crude for Arabian Light marker from $7.113 to $8.260/b and raised the average government take on all crudes from $9.41 to $9.74/b. Government crude, which was 60 percent of production in most cases, was to be sold at 93-94.8 percent of posted prices.[28] Saudi Arabia agreed that in view of what it perceived as excessive levels of profits made by the oil companies, the increases in royalty and tax rates were justified, and advocated that these increases be accompanied by a reduction in posted prices in order to give consumers some respite. The Saudis postponed the incorporation of July 1 and September 1 increases in taxes and royalties pending the outcome of the ongoing participation negotiations with Aramco.

The participation negotiations, however, were not making satisfactory progress from the Saudi standpoint. The chief point of contention was the Saudi insistence that the companies purchase their liftings at open market prices rather than at a preferential price. In view of a continued impasse, Saudi Arabia made two crucial decisions that were bound to produce desirable results for her. First, the government announced on September 27 that it was going to implement fully the OPEC increases of July 1 and September 1. Second, at a meeting of the Gulf states in Abu Dhabi, Saudi Arabia proposed a formula aimed at reducing the posted price of Arabian Light from $11.651 to $11.251/b and increasing tax rates from 65.75 to 85 percent and royalties from 16.67 to 20 percent on participation crudes of all companies. If implemented, the Saudi proposal would have increased government take and made crude oil somewhat cheaper for customers of NOCs and more expensive for the international majors. The oil states were reportedly considering the pricing formula set forth by Iran, which proposed a unified price agreed to by all oil states, with additional charges for oil quality and geographic location. This mechanism proposed a basic price for crude that would last one year at around $10.35/b.

The OPEC Ministerial Conference in Vienna, December 12-13, 1974, decided to combine the Saudi and Iranian proposals. As a result of this combination, the direct sale price of government-marketed crude was to be $10.46/b, and the government take was to be raised to about $10.12/b on Arabian Light marker crude. By adopting the new pricing system, which was to be in effect from January 1 through September 30, 1975, OPEC not only eliminated the posted price system but also attempted to fix the oil companies' margin on oil production to 22¢/b. Saudi Arabia, the UAE, and Qatar were to announce their official position on this decision pending the outcome of the on going negotiations on participation.[29]

Table 11. Oil Production Losses by OPEC Members, 1974-1975 (1,000 b/d)

Producer	1974	1975	Volume Change	% of Change
Algeria	1,008.6	1,020.3	−11.7	−1.2
Ecuador	177.0	160.9	−16.1	−10.4
Gabon	201.5	223.0	+21.5	+9.6
Indonesia	1,374.5	1,306.5	−68.0	−5.2
Iran	6,021.6	5,350.1	−671.5	−12.6
Iraq	1,970.6	2,261.7	+291.1	+12.9
Kuwait	2,546.1	2,084.2	−461.9	−22.2
Libya	1,521.3	1,479.8	−41.5	−2.8
Nigeria	2,255.0	1,783.2	−471.8	−26.5
Qatar	518.4	437.6	−80.8	−18.5
Saudi Arabia	8,479.7	7,075.4	−1,404.3	−19.9
UAE	1,678.6	1,663.8	−14.8	−0.9
Venezuela	2,976.3	2,346.2	−630.1	−26.9
Total	30,729.2	27,192.7	−3,536.5	−13.1

Source: Figures for 1974 and 1975 production are from *Annual Statistical Bulletin*, 1975 (Vienna: OPEC).

Considering the magnitude of the price increases introduced by OPEC in October and December 1973, the negative effects of these decisions were not felt as much as they should have been by the oil states in 1974. As shown in Table 10, a comparison of 1973 and 1974 total OPEC production shows that the demand for OPEC oil was reduced by only 0.8 percent in the latter year. Total production took a substantial plunge in 1975, from 30.73 million to 27.19 million b/d, a reduction of 13.1 percent. Levels of production cutbacks by individual OPEC members from 1974 to 1975 are shown in Table 11. For the industrial consumers, the effect of these price escalations was noticeable changes in economic activity. For instance, the growth of real gross national product (GNP) of the twenty-four-member OECD, which had averaged about 5.4 percent annually from 1959 through 1972, fell to nearly zero.[30] While it would be incorrect to put the burden of blame for this on the quantum price increases introduced by OPEC, it can be prudently stated that these increases made a significant contribution to the economic slowdown of 1974. By the end of 1974, a variety of conservation-oriented measures were adopted by the Western industrial countries. These measures included lower thermostat settings, reductions in motor vehicle speed limits, rationalization of airline schedules, and a general attention to conserving fuels. By February 1975, these conservation measures were reported to have resulted in the following percentages of energy savings: United States, 4.9 percent; France, 5.7 percent; West

Germany, 14.3 percent; Italy, 3.7 percent; United Kingdom, 9.5 percent; Belgium, 19.4 percent; Netherlands, 14.2 percent; and Denmark, 18.6 percent.[31] In addition, the newly-created International Energy Agency (IEA) announced that its eighteen member states had decided to reduce their expected oil imports by 2 million barrels or about 10 percent, by the end of 1975. It was calculated that this cut actually amounted to about 6 million b/d if the imports were compared to what they might have been if OAPEC had not imposed an oil embargo and if OPEC countries had not quadrupled the price of oil.

In summary, at least three broad effects of the unilateral price increases may be identified. First, this unilateral capability did not protect the oil states from the erosive effects of international inflation. In fact, except for introducing minor upward adjustments in royalties and taxes as feeble attempts to reduce the profits of the oil companies, OPEC members were unable to bring about major price escalations during 1974 and 1975. Second, the elimination of the oil companies from pricing decisions by no means made these decisions a smooth process for OPEC. On the contrary, the moderating role of the oil companies was permanently taken over by Saudi Arabia. Consequently, OPEC price discussions since 1974 have often been marked by acrimonious debates and in-group tensions. Third, for the oil consumers, the unilateral price increases not only worsened the effects of the economic slump of 1974-1975 but also resulted in lopsided deficits in their current accounts as billions of dollars were taken out of their economies in the form of OPEC tax. For the oil states, the huge oil taxes enabled some large producers to introduce substantial production cutbacks and thereby minimize the downward pressure on prices of crude oil resulting from the economic slump.

The Attempt to Create NIEO OPEC's price decisions of October 1 and 16 and December 23, 1973, created considerable chaos in the world economy. Although the oil states succeeded in quadrupling the prices of crude oil, they remained wary that their resultant financial gains would be only temporary as international inflation continued to escalate. To assure long-term prosperity, they needed cooperation from the industrial consumers. Cooperation, however, is based upon a certain amount of mutual trust among the parties and, most importantly, should be evolved to promote congruity of mutual interests. The industrial consuming nations recognized the importance of cooperation, but the basis for mutual trust was completely lacking. They were resentful of what they considered OPEC's complete disregard of the implications of their unrealistic increases for the non-Communist economy. Most oil states, on the contrary, convinced that their price increases were economically justified, were threatening to introduce similar increases again if international inflation remained unchecked. And the responsibility for slowing the inflationary rates, as the oil states perceived it, rested primarily with the industrial consumers, not with

OPEC. Under such circumstances, OPEC set out to initiate serious dialogues with the industrial consumers, the purpose of which was to correct what they and their Third World allies regarded as an asymmetrical economic relationship between the industrial and the nonindustrial nations.

At its December 1974 meeting in Vienna, OPEC decided to convene a joint meeting of its oil, finance, and foreign ministers in Algiers on January 24, 1975. Its purpose was to prepare an agenda for the summit conference of OPEC countries, which was originally proposed by President Houari Boumedienne of Algeria. The Algiers meeting was also to give general consideration to economic and monetary issues for the purpose of developing OPEC's position for a French-sponsored trilateral conference composed of OPEC and industrial and nonindustrial countries, to be held from February 21 to March 8.[32]

At the conference in Algiers, January 24-26, President Boumedienne set the tone of OPEC's future strategy by advocating the need for and importance of dialogue between industrial countries and producers of primary commodities, including oil. A hallmark of this conference was its moderate tone. Various major proposals discussed were derived from the "Declaration and Program of Action Concerning the Establishment of a New International Economic Order," adopted by the Sixth Special Session of the United Nations General Assembly in April 1974, which was devoted to the problems of raw materials and development. Some of the significant proposals of OPEC's Algiers conference are worth mentioning. The oil states wanted the Western countries to perform the following actions: (1) to end "confrontation tactics" and initiate a "constructive dialogue on the basis of equality" with the nonindustrial countries; (2) to establish an "equitable relationship" between oil prices and prices of goods and services imported by OPEC; (3) to "boost the economic, social and cultural development of the Third World, by facilitating, among other things, the transfer of technology and funds and opening up their markets to the products of Third World industries"; (4) to eliminate restrictions "on the free employment of liquid assets by the OPEC countries"; (5) to develop "appropriate pricing schemes for other Third World raw materials apart from oil"; and (6) to reform the international monetary system and "to allow for active participation in decision-making by the Third World."

As a quid pro quo, OPEC promised three measures. First, the organization promised not to increase the prices of oil in real terms between 1975 and 1980. The price schedule of January 1 through September 30, 1975, was to be extended for the remainder of that year. For the following years, the prices were to rise partially to reflect the rates of inflation prevailing in 1976 and 1977, and then they were to be adjusted to compensate for the full rate of inflation between 1978 and 1980. Implementation of this proposal would have caused a decline in the real value of oil between 1975 and 1977. The second measure was OPEC's guarantee to provide to the world market all the oil needed. Third, OPEC proposed a plan for recycling petrodollars which included the granting of credits for the purchase of oil to those OECD countries encountering balance-of-payment problems.[33] These proposals were enticing enough to deepen further cleavages in the already divided ranks of the industrial consumers. For example, the European countries and Japan were far more concerned about the availability of petrodollars for future use and the security of supplies than was the United States, whose total oil imports at that time remained around 35 percent, of which the Arab oil accounted for a little over 10 percent.

While OPEC was searching for ways to calm the price situation and thereby assure its long-term prosperity, the uncertainties of the international oil market were beginning to look more ominous for the oil states. There were reports of "cracks" in that organization's solidarity amid instances of major production cutbacks and price reductions.[34] Saudi Arabia announced a production slash of about 1 million b/d for January, from 8.5 million to 7.6 million b/d. By May the Saudi production fell from 6.3 million to slightly under 5.7 million b/d. Reportedly, this was the lowest level of production since July 1972, when the output averaged 5.2 million b/d. Considering that Aramco's production capacity at that time remained at 11.2 million b/d, this was a huge drop.[35] Production in Libya was also reported to be hitting its lowest levels in 10 years, about 1 million b/d. Libya was also involved in price reductions by slashing its differential costs (premiums added to the price of oil to capitalize on proximity to market and/or quality of oil) between 30 and 50¢/b. This reduction was a significant depression in price, considering that the buy-back price of Libyan crude reached $16/b in early 1974. It should be pointed out, however, that even in the range of $11 to $11.20/b, the price of Libyan buy-back crude was still higher than the price of the Arabian Light marker crude, which then prevailed at $10.46/b. Algeria, Nigeria, and Abu Dhabi were also reportedly involved in reducing certain premiums.

The oil states were expected to take remedial measures against the sagging world demand along with other important issues, such as the agenda for the forthcoming summit between the oil-producing and oil-consuming nations, and a reconsideration of the in-place price freeze in view of the declining value of the U.S. dollar. But OPEC's Extraordinary Meeting in Vienna, February 25-27, 1975, took no action to improve OPEC's grip on the oil market by incorporating coordinated production programing. At Saudi insistence, the oil states agreed to continue the price freeze. This Saudi action was reportedly based upon an understanding reached with the United States during the pevious week. As a result of this understanding, the United States was reported to have extended its support to the conference between oil-consuming and oil-producing countries on the price of oil and other raw materials. In return, Saudi Arabia was understood to have promised to support the continuation of a freeze on the price of oil for 1975 and to persuade other oil states to introduce modest reduction. The conferees instructed oil, finance, and foreign ministers of OPEC countries to continue drafting the "Solemn Declaration" of the program of action for the March 4 meeting, which was to be submitted for approval by heads of oil states as a general framework for negotiations with the consuming nations.

OPEC's summit conference at Algiers, March 4-6, 1975, was a special occasion for the petroleum-exporting countries, for it provided OPEC with true international status: from an organization whose original *raison d'être* was to coordinate and unify the petroleum policies of member countries and to devise ways and means of insuring the stabilization of prices in the international oil market, it now became a body whose primary concerns also included the establishment of a New International Economic Order. Even though the agendas of previous OPEC meetings included consideration of these issues, the summit at Algiers was a full-fledged occasion to consider issues of global dimension, not just those topics that were of primary concern to the oil states. The "Solemn Declaration" issued at the conclusion of this conference was the product of a compromise between radically divergent outlooks of OPEC members. This statement also set the tone as well as the framework of this conference and the impending conference between the producing and consuming nations. The primacy and centrality of oil prices in the dialogue between the two groups of nation-states was emphasized by the host of this meeting, President Boumedienne, when he declared: "If prices have to be frozen, we will freeze them; if they must be decreased, we will decrease them provided, however, that the developed countries make similar and simultaneous effort in return—with each contributing according to their means and responsibilities to the reorganization of the world economy and the establishment of the stability required for development and prosperity."[36]

The conferees had strong disagreements on three crucial subjects: production programing, the indexing of oil prices, and the use of surplus funds. A sizable majority of members, spearheaded by Algeria, Iraq, and Libya and joined by Iran, Venezuela, and Nigeria, labeled the prevailing surplus of oil "artificial" and advocated immediate adoption of production programming. Saudi Arabia, on the contrary, reiterated its continuing opposition to this mechanism. The indexing of oil prices was gaining considerable support, especially among the price hawks. Saudi Arabia did not disagree with the principle of indexing to compensate for inflation and currency fluctuations. But the points of contention were when and to what extent indexing should be introduced. The price hawks, as an expression of their growing concern over the erosion of their purchasing power, wanted to introduce full indexing immediately. Saudi Arabia, in contrast, was pressing for a modest reduction in the basic level of crude prices beyond the end of 1975, if possible, before agreeing to indexing. The incorporation of the Saudi proposal would have caused further erosion of prices in real terms.

On the employment of surplus funds, the contrast between the attitudes of the conferees representing the capital-surplus and the capital-deficit oil states was remarkably similar to the contrast between the developed and the developing countries on aid and trade. The capital-deficit states advocated that surplus funds that are channeled to the developed and Third World countries remain under the administrative control of a centralized OPEC body. The capital-surplus oil states (Saudi Arabia, Kuwait, Iran, and the UAE), which wanted to maintain full control over their contributions to such arrangements, categorically rejected any supranational OPEC control of the surplus funds. A majority of the participating states agreed to support the Saudi measure, which proposed that OPEC aid to the developing countries be channeled through the already-established development funds in various OPEC countries and that the joint use of OPEC surplus funds for recycling and other purposes be studied by the finance ministers of capital-surplus oil states only. The conference also resolved the cumbersome question of OPEC representation at the preliminary producer-consumer meeting. Algeria, Saudi Arabia, Iran, and Venezuela were selected to represent the OPEC bloc. The non-oil-producing countries of the Third World were to be represented by India, Brazil, and Zaire; the Western industrial consumers were to be represented by the United States, the member states of the European Economic Community (EEC), and Japan.

Representatives of the non-oil-producing countries of the Third World and OPEC, in a meeting on April 5, decided that the Third World countries should be represented as one bloc at the forthcoming Conference on International Economic Cooperation (CIEC). They also agreed that their agenda should be based on the "Solemn Declaration" developed at the January 31 and March 7 OPEC meetings and the 1974 special session of the United Nations General Assembly on Raw Materials and Development. The preparatory meeting between the representatives of the Third World and the Western industrial countries on April 7 collapsed, however. The participants were too obsessed with the notion of limiting the focus of the CIEC to promote their own economic priorities and interests. For instance, the industrial consumers wanted to use the conference to find avenues for reducing oil prices. OPEC members wanted a discussion of mechanisms for indexing oil prices. The industrial consumers expressed their willingness to initiate negotiations on indexing provided OPEC members agreed to lower the base price. For the oil states, lowering oil prices was too high a price to pay for negotiating the issue of indexing. Concerned that the Western industrial countries might maneuver the conference into spending an inordinate amount of time and assigning undue priority to discussion of energy, the non-oil-producing countries of the Third World insisted that the proposed conference be equally concerned with other problems of the developing countries. The agenda of the Third World countries included: oil and other raw materials, including foodstuffs; revalorization and protection of purchasing power of the export earnings of the developing countries (which involved the indexing of energy prices and at least some raw material prices); financial cooperation; assistance to the most seriously affected countries; reform of the international monetary system; promotion of nondiscriminatory and nonreciprocal preferences for the developing countries, access to the markets of the industrial countries, and reindustrialization. The developed countries, on the other hand, maintained that the discussion of broader economic issues should be only peripheral to the central problem of energy. The preparatory meeting, under such a charged political climate, was destined to fail. The degree of polarization of the participants may best be indicated by the fact that they could not even agree on what the conference should be called.[37]

With the prospects of the dialogue between the consuming and producing states undergoing a period of impasse and uncertainty, OPEC once again was faced with the obdurate task of taking immediate measures regarding indexing and the declining value of the U.S. dollar, a subject of considerable concern since September 1974. OPEC's price escalations of the previous year had made a substantial contribution to this decline. For the United States, the estimated dollar outflow attributable to oil price increases for 1974 was $25 billion. The Department of Commerce estimated that the soaring U.S. oil bill sent the nation's balance of payments into a $900 million deficit for the first half of 1974 and warned that continued deficits would lead to further weakening of the dollar compared to that of other currencies.[38] Prior to OPEC's next scheduled meeting in Gabon, the prospects of consumer-producer dialogue were given a minor boost by Secretary of State Henry Kissinger's statement expressing U.S. willingness to reconvene the preparatory meeting in the format previously agreed to. The Third World bloc and the Western consumers made virtually no progress on issues that had kept them far apart at the last preparatory meeting. Evidence of continued polarization of the parties is the fact that the United States maintained its opposition to the issue of indexing, a topic that was very popular with the oil states. In addition, the United States continued to insist that the emphasis of the meeting be on energy. Kissinger's statement was viewed as a tactic to postpone further price increases, which the oil states were expected to consider in their forthcoming meeting.

Kissinger's statement created but few ripples at the next OPEC conference at Libreville, Gabon, June 9-11, 1975, where the oil states decided to continue the price freeze that had been in effect since December 1974 until the end of September 1975. In view of the continued weakening of the dollar, the oil states were expected to switch to Special Drawing Rights (SDRs)[39] as a unit of account for oil transactions. But this decision was put into abeyance as a result of Saudi Arabia's insistence that such a switch must be made without any actual price increases in dollar terms. The oil states also decided to pigeonhole the issues of price and freight differentials.[40]

If the Libreville conference was marked by a lack of decisiveness on the part of the oil states, then the following meeting in Vienna, September 24-27, 1975, was noted for acrimonious debates between Iran and Saudi Arabia, two countries that were a world apart on the nature of future price increases and whose disagreement on this issue was further exacerbated by the renewal of traditional rivalry for the leadership of OPEC. In a preconference meeting the Shah made it clear to Yamani of Saudi Arabia that Iran would in no way accept a price increase of less than 15 percent. This was ostensibly a major concession on the part of Iran, whose earlier public posturing had been for a 30-35 percent increase. Saudi Arabia would have preferred a further continuation of the price freeze, but sounded somewhat receptive to a 5 percent increase.

At the Vienna conference, the proposed Iranian rate of increase was well received by such varied actors as Venezuela, Iraq, Nigeria, Indonesia, Qatar, Gabon, and Ecuador. The Iranian position was substantiated by reports by two OPEC groups. The Economic Commission reported that the rate of inflation for the first nine months of 1975 stayed at 20 percent. The OPEC committee of experts estimated that the imported inflation for the same period was 28 percent. After considerable bickering, the conferees, under the Saudi threat of unilaterally lowered prices and increased production, agreed on a 10 percent increase. This decision, which was to be in effect from October 1, 1975, through June 30, 1976, raised the price of Arabian Light marker crude from $10.46 to 11.51/b and escalated the overall cost to oil consumers by more than $10 billion. The oil states were unable to take any decisive action on switching to SDRs as a unit of account for oil transactions and on differentials during the remainder of 1975.

All in all, 1975 produced mixed results for OPEC. Despite the economic downturn emanating from the recession of 1974-1975, OPEC retained its overall grip on the oil market because the OPEC "tax" earnings in 1974 enabled its members to earn $100 billion, all but $5 billion of which was received in payment for oil exports. It was estimated that about $35 billion of OPEC's earnings was to be spent for imports of goods and services. After subtracting grants to Third World countries and adding income on their investments, OPEC members were left with roughly $60 billion as current account surplus in 1974.[41] Of this $60 billion, about $26 billion belonged to the Arab oil states with limited absorptive capacity. These included Saudi Arabia, Kuwait, Libya, and the Persian Gulf sheikhdoms.[42] Table 11 shows that financially these states were capable of absorbing large production cutbacks without any hardship. OPEC's ability to sustain production cutbacks neither ameliorated that organization's concern about inflation nor slowed down its quest for mechanisms to keep members' purchasing power from being continually eroded.

On the so-called North-South dialogue, oil power failed to produce significant progress. In its zeal to create a new international economic order, the Third

World bloc lost touch with reality. Oil power was a phenomenon of enormous potential, but it could not be expected to produce miracles. And the problems involved in the North-South dialogue were enormously intricate and needed an equally enormous degree of patience and realism, especially on the part of the Third World bloc. These actors behaved quite differently and, despite a number of negotiating attempts made by the consuming and producing nations between 1975 and 1977, the prospects of progress on these issues remained as elusive as ever.

For the industrial consumers, the OPEC price escalations emerged in the form of $40 billion deficits in current accounts for 1974, but they dipped down to $6 billion for 1975, largely as a result of the tremendous economic slump caused by the recession of 1974-1975.[43] As a result of the Arab oil embargo and OPEC price escalations, the relationships between OPEC and the OECD countries and among the industrial countries themselves went through some interesting phases.

Disarray, Confrontation, and Cooperation The industrial consuming countries were disgruntled by the intermittent leapfrogging in the price of oil from 1971 through 1973. The quantum leaps of October and December 1973 and the Arab oil embargo, however, sent shockwaves through their ranks. Their recovery from the "shocks of 1973" was long and drawn out. The oil embargo resulted in worldwide perception of an energy crisis. The consuming nations and the international oil companies appeared overwhelmed by the remarkable show of unity demonstrated by the Arab oil states under the aegis of OAPEC.

As an initial response to the oil embargo, American decisionmakers toyed with several policy options. For instance, a counterembargo of shipments of food and manufactured goods to the Arab oil states was briefly considered. But in the absence of multilateral cooperation, such a move appeared futile and was thus rejected. Similarly, the option of curtailing arms shipments to such traditional American arms buyers as Saudi Arabia and Kuwait was also set aside, since these states could have purchased the needed weaponry either from the Europeans or from the Soviet Union. The option of creating an organization of oil importing countries appeared promising largely because it was based upon realism, and the United States was certain that, in due time, the OECD allies might be persuaded to join such an organization. On March 31, 1973, Senator Thomas McIntyre called for collective actions by the consuming states, including the United States, to counter the power of the profit-oriented oil companies and producing states. Eight other senators on April 6 repeated this call and urged President Nixon to call a conference of consuming states to explore creation of a counter-organization. On April 10, Deputy Secretary William Simon stated that a joint approach by the consuming nations might be the best protection from the

Table 12. Estimated Oil Imports by Source, 1973

Importing Country	Oil Imports as % of Energy Consumption	Total Oil Imports (1,000 b/d)	Source of Imports (1,000 b/d)													
			Arab Producers								Other Producers					
			Saudi Arabia	Kuwait	Libya	Iraq	Abu Dhabi	Algeria	Other	Total Arab	Iran	Venezuela	Indonesia	Canada	Nigeria	Other
United States	17	6,200	590	160	350	50	160	140	140	1,590	420	1840	250	1,100	550	450
Percent			9.5	2.6	5.6	0.8	2.6	2.3	2.3	25.6	6.8	29.7	4.0	17.7	8.9	7.3
Canada	27	1,000	80	[a]	40	20	60		20	220	180	470	[a]		80	50
Percent			8.0	[a]	4.0	2.0	6.0		2.0	22.0	18.0	47.0	[a]		8.0	5.0
United Kingdom	52	2,330	550	400	240	40	50	50	150	1,480	460	80	[a]		180	130
Percent			23.6	17.2	10.3	1.7	2.1	2.1	6.4	63.5	19.7	3.4	[a]		7.7	5.6
West Germany	42	2,200	480	90	550	30	110	280	70	1,610	270	40	[a]		200	80
Percent			21.8	4.1	25.0	1.4	5.0	12.7	3.2	73.2	12.3	1.8	[a]		9.1	3.6
Italy	93	2,410	630	200	460	390			250	1,930	330	20			10	120
Percent			26.1	8.3	19.1	16.2			10.4	80.1	13.7	0.8			0.4	5.0
France	77	2,690	610	310	150	350	290	230	130	2,070	220	30			250	120
Percent			22.7	11.5	5.6	13.0	10.8	8.6	4.8	77.0	8.2	1.1			9.3	4.5
Netherlands	57	2,070	690	380	60	10	80	20	140	1,380	440	10			220	20
Percent			33.3	18.4	2.9	0.5	3.9	1.0	6.8	66.7	21.3	0.5			10.6	1.0
Norway	47	140	20		3	3	7	2	29	64	55	8			12	1
Percent			14.3		2.1	2.1	5.0	1.4	20.7	45.7	39.3	5.7			8.6	0.7
Belgium/ Luxembourg	60[b]	720	290	120	30	30	10	50	20	550	100	20			30	20
Percent			40.3	16.6	4.2	4.2	1.4	6.9	2.8	76.4	13.8	2.8			4.2	2.8

Importing Country	Oil Imports as % of Energy Consumption	Total Oil Imports (1,000 b/d)	Source of Imports (1,000 b/d)													
			Arab Producers								Other Producers					
			Saudi Arabia	Kuwait	Libya	Iraq	Abu Dhabi	Algeria	Other	Total Arab	Iran	Venezuela	Indonesia	Canada	Nigeria	Other
Denmark	94[b]	200	62	32	3				36	133	39					28
Percent			31.0	16.0	1.5				18.0	66.5	19.5					14.0
Turkey	41[b]	160	66			29			54	149	10					
Percent			41.3			18.1			33.7	93.0	6.3					
Greece	67[b]	150	72			46		7	9	134	2					14
Percent			48.0			30.7		4.7	6.0	89.4	1.3					9.3
Portugal	80[b]	90	25			36			11	72	7					11
Percent			27.8			40.0			12.2	80.0	7.8					12.2
Spain	64[b]	1,000	470	90	40	50		110	60	820	120	40			10	10
Percent			47.0	9.0	4.0	5.0		11.0	6.0	82.0	12.0	4.0			1.0	1.0

Source: U.S. Congress, House, Hearings before the Subcommittees on Europe and on the Near East and South Asia of the Committee on Foreign Affairs, *United States-Europe Relations and the 1973 Middle East War,* 93 Cong., 1 and 2 sess. (Washington, D.C.: GPO, 1974)

Oil import figures for several countries, particularly the Netherlands, Italy, and Spain, are higher than domestic consumption of imported oil because these countries are signficant exporters of refined petroleum products.

[a]Negligible.

[b]1972 data; 1973 data not available.

growing power of the Middle Eastern oil states. Soon after his statement, U.S. officials initiated consultations with the allies to draft plans for an organization of importing countries whose principal members were to be the United States, the members of the European Common Market or European Economic Community, and Japan, with other countries free to join. The stated purpose was to devise an international system of allocation of available oil if chronic shortages should occur, and to avoid an international price war in which importing nations would bid up prices. In reality, State Department officials hoped that a "buyers' cartel" (even though they reportedly eschewed that phrase) might offset the price-raising ability of OPEC.[44]

The problem with such an inherently confrontational approach was that the willingness of American allies to rally around it would be determined by the nature of their dependence on foreign oil and, more importantly, on OPEC's reactions to the American maneuvers to form a consumer bloc. The patterns of energy supplies for the United States, Western Europe, and Japan were radically different. As shown in Table 12, the United States, whose energy imports in 1973 were only around 17 percent of its total energy consumption, was less dependent on foreign oil than any of its allies, while European dependence varied from a high of 94 percent for Denmark to a low of 42 percent for West Germany. The Canadian imports of 27 percent were insignificant, since Canada was an oil-producing country. Then there was a political dimension which was to play an overriding role in determining allied behavior. As the allies perceived it, imposition of the Arab oil embargo was a more or less direct outcome of the extremely pro-Israeli posture maintained by the United States. By becoming a party to American attempts to create a consumer bloc, the allies were afraid they not only might apear to support American policies toward Israel, but also might be viewed by the Arab states as parties to the confrontational politics practiced by the United States. Thus the Common Market members were not expected to join the U.S. bandwagon in confronting either OPEC or the OAPEC countries. Furthermore, the overall negative reception by the oil states to American efforts to create an alliance were an additional deterrent to the development of cohesive behavior among the industrial consuming nations.

The first formal reaction from OPEC to the American endeavors came in a communiqué issued at the end of an Extraordinary Meeting in March 1973. OPEC warned the consumers against forming a cartel of their own and stated that such measures would be quite contradictory to orderly international trade. The oil states were only too well aware of how potentially formidable a consumers' cartel could be. After all, the power of OPEC rested with its ability to act as a price-fixing cartel, a fact that was abundantly clear at the time.[45] Zaki Yamani of Saudi Arabia used a harsher tone than the one expressed in the March communiqué when he stated that any attempts to form a consumer cartel could lead to an oil war, the consequences of which would be damaging to the industrialized states. In general, OPEC members were more than satisfied with the general disarray caused among the Western industrial countries as a result of their warnings.

The shocks of 1973 caught the industrial consumers quite unprepared. Their initial response was to develop elaborate energy policies, and the issue of the oil supply was no longer the sole prerogative of the multinational oil companies. The worldwide operation of these companies remained intact, however; only their responsiveness to the governments increased with the growing activism of the governments in energy affairs. An important aspect of increased governmental activism was a willingness to acquire access to oil at any cost. This led to the development of direct government-to-government (bilateral) deals between the oil states and industrial consumers. Essentially such deals involved exchanges of a whole range of goods and services, including agricultural development schemes, blueprints of industrialization, petrochemical and refining equipment, and sophisticated arms, in exchange for oil. The industrial consuming nations, assuming that the oil states would stand by their commitments, viewed these bilateral deals as added security. The more dependent an industrial country was on foreign oil, the more actively and even aggressively it sought bilateral deals with the oil states. Japan, a nation almost totally dependent on Middle Eastern oil supplies, declared that it no longer considered petroleum a mere commercial item but regarded it as linked solely to Japanese politics and national interests. Accordingly, that country resolved to develop a bilateral approach to petroleum affairs, based on diplomacy. In addition, the Japanese emphasized the element of cooperation with OPEC members rather than competition, lest their attitude be interpreted as confrontation by the oil states.[46]

The shocks of 1973 also created consternation and disarray among the Common Market countries. The Arab oil embargo not only made these countries acutely aware of their vulnerability, but also dealt the notion of European unity a severe blow. A glaring example of this disarray was the European response to the effect of the embargo on the Netherlands. Along with the United States, the Netherlands was one of the primary targets of the Arab oil embargo because of its alleged pro-Israel policies. When the Netherlands felt that it would be drastically short of oil, it requested oil-sharing arrangements with its European allies. The Dutch request, however, clashed head on with the desire of France and Great Britain to avoid a showdown with Arab countries. This situation resulted in a "vulnerble state of indecision" among European governments, combined with a tacit preference for delegating to the international oil companies the responsibility of allocating scarce oil supplies among the various countries.[47] When the original shocks subsided somewhat, the Common Market countries sought to develop common energy policies within their own ranks and closer contacts with the major oil-producing countries as well as with major oil-consuming nations such as the United States and Japan. While actively seeking cooperative avenues on energy matters among consuming nations, the Common Market countries continued to reject the U.S. suggestions to form an association of consuming nations which would function as a bloc in all dealings with OPEC states.[48]

OPEC members had expressed a special interest in access to the technology of the industrial consumers in the communiqué of March 1973. The precedence of a bilateral deal had been established long before the oil embargo was imposed, when Japan signed an agreement to buy 70,000 b/d of Saudi participation crude at well above the market price.[49] This was perhaps one of the first arrangements in what later emerged as a pattern of bilateral deals between the industrial consuming and oil-producing nations.

In July 1973, OPEC made a significant announcement that its members might be compelled to buy manufactured goods as part of the payment for their oil. This statement gathered momentum for bilateral deals. In January 1974, in a post-meeting communiqué, OPEC reiterated its preference for bilateral arrangements by announcing that its member states wanted the consuming nations to approach them on future bilateral deals instead of issuing an invitation for consultations. This was an obvious rebuff to the ongoing U.S. endeavors to launch an "energy action group." At this meeting Saudi Arabia announced the conclusion of a bilateral "arms for oil" agreement with France. Even though Yamani tried to minimize the significance of this agreement by labeling it "small," he added that its total cost amounted to $2 billion.

As the industrial consuming countries continued their scramble to conclude bilateral deals with the oil states, it became obvious that a new, unfamiliar, and unprecedented system of international power relations was emerging, one in which the oil states would be given considerable importance by the leading industrial consuming countries. It was equally obvious that to a majority of the consuming nations, this new power arrangement did not appear to be systematic, orderly, or legitimate. It was felt in some quarters of the industrialized world that this effort to woo the oil producers might prove perilous in two ways. First, the eagerness to work unilateral deals might tempt individual consumers to outbid each other and thereby push the international prices still higher. Second, in the absence of deliberate cooperation on oil affairs, the individual consuming nations might become susceptible to blackmail at some stage by one or more oil states.[50]

In response to such sentiments and as a continuation of its sustained search for long-range solutions to price escalations and oil shortages, the United States, which strongly opposed bilateral deals, organized the much-heralded energy conference in Washington on February 11, 1974. The oil states remained suspicious of this gathering of consumers. Their sentiments perhaps were best expressed by an Arab League envoy when he observed that to confine the meeting to a select group of industrialized nations was tantamount to affirming a continued economic hierarchy.

Presumably because of the negative posture of the oil states vis-á-vis the energy conference and because of the continuing oil embargo, the United States and its allies were unable to find much common ground. The differences between these nations remained vivid even in the few days preceding the Washington conference. In holding this conference, the United States called for a comprehensive cooperative program to meet the energy crisis. The Common Market countries, on the contrary, in a conference on February 5 adopted a joint position calling for a general review of the energy crisis and of possible areas of cooperation—an affirmation of a narrower goal.

The Washington energy conference was the beginning of systematic cooperation by the Western industrialized countries, even though France and Japan remained opposed to confronting the OPEC states. The specifics of an oil-sharing plan in the wake of an emergency were one of the most serious issues of contention between the industrial consumers. At the conference the United States made a major concession by agreeing to share "indigenous" oil supplies (i.e., oil produced within the United States) in times of emergency and severe shortages. This was a change from its earlier position to share only oil moving in international trade (its oil imports). The energy conference also established a twelve-nation Energy Coordination Group (ECG) charged with developing international cooperative arrangements among the industrial consumers.[51] France, because of its persistent opposition to forming any cooperative agreement that might intimidate the oil states, decided not to join the ECG.

Numerous meetings between ECG member countries produced a ten-year blueprint for a new "international energy program" (IEP). An International Energy Agency (IEA),[52] which was to replace the Energy Coordinating Group, was to be established as a separate entity to administer this program within the OECD. The IEP contained short-term programs as well as long-range policies. The short-term programs were aimed at protecting oil consumers, discouraging them from bidding competitively for available oil supplies, establishing stockpiling requirements and stockpile drawdown rules for all participants, and spelling out a variety of measures aimed at mandatory demand constraints on specified levels of supply shortfalls. The purpose of long-range policies was threefold: (1) to develop among the oil importers long-term cooperative mechanisms aimed at reducing dependency on imported oil; (2) to create an oil market information system aimed at improving knowledge of the functioning of the world oil market and to create a framework for consultations with individual oil companies; and (3) to develop a framework for coordinating relationships among industrial consumers, oil exporters, and the Third World countries, and to elaborate a common strategy for the eventual dialogue with the oil states.

The international energy program included several short-term programs:

emergency stockpiling, intended to serve as a deterrent to future supply interruptions; emergency demand restraint, aimed at substantially improving participants' ability to withstand the economic impact of an embargo; and emergency oil-sharing, intended to serve as assurance that all member countries come to the assistance of any partner whose oil imports were dropped by more than 7 percent as a result of selective embargo. Long-range policies included energy conservation policies aimed at reducing excessive dependence on imported oil; long-range demand restraint (reduction of the oil imports of participating countries as a whole by 2 million b/d by the end of 1975 and possibly by 4 million b/d by the end of 1977); alternate energy development (establishment of a $25 billion financial safety net to serve as the lender of last resort); and creation of a coordinated system of cooperation in the accelerated development of new sources of energy comprised of (1) an agreement to encourage and safeguard investment in most conventional energy sources through establishment of a common minimum price below which member states would not allow imported oil to be sold within their economies; (2) a framework of cooperation to provide specific incentives to investment in higher-cost energy on a project-by-project basis; and (3) cooperation in energy research and development, including the pooling of national programs in selected projects.[53]

With the IEP firmly established, the institutionalized process of cooperation among the Western consuming nations was well on its way. The cooperation between oil-consuming and oil-producing countries, however, remained as elusive as ever. The United States never really abandoned its confrontational posture toward OPEC. In the aftermath of the energy conference, the United States was obliged to tone down its rhetoric because, instead of persuading the industrial consumers to the American point of view, it appeared to be escalating the level of paranoia in France and Japan. These countries were in no mood to try to intimidate OPEC. In fact, France was the only industrial country which opted not to join the IEA. The United states, as a positive gesture toward France, gave a reluctant endorsement to the French-proposed North-South conference. U.S. participation in this conference was never really accompanied by a genuine desire on the part of this country to work toward meaningful progress on the intractable issues of the North-South dialogue. No other industrial country, with the exception of France, was really interested in keeping these dialogues alive.

The oil states bear no less blame for the lack of cooperation. OPEC's price increases of 1973 created an environment nonconducive to cooperative endeavors such as the North-South dialogue. In fact, such an environment enhanced the possibility that the Third World producers of copper, bauxite, rubber, and other raw materials might be able to exercise maximum leverage through withholding supplies altogether from the industrial consuming nations or might use their monopoly power to charge exorbitant prices for their raw materials.[54] It was only rational for the industrial countries to nip such potentialities in the bud.

Attempting to deal with OPEC was difficult enough; the notion of two or more cartels was nightmarish indeed. An unfortunate consequence of such an environment was that the North-South dialogue suffered a serious setback from which it did not recover. Toward the end of 1975, when the industrial countries were gradually pulling themselves out of the great recession of 1974-1975, the chances of the economies of the non-oil-producing developing countries recovering from this recession did not appear promising. OPEC countries were still grappling with the issue of pricing and inflation, and the United States was still obsessed with dismantling OPEC.

The Participation Agreements

The participation agreements were the most noteworthy achievements of OPEC in the sense that they brought about qualitative changes in the historical relationship between the oil-producing countries and the oil industry. OPEC's first formal position on the subject was spelled out in a major announcement, the "Declaratory Statement of Petroleum Policy in Member Countries," in June 1968.[55] No substantive action was taken until July 1971, however, except for a number of in-depth studies by the OPEC secretariat. At the Twenty-Fourth Conference, at Vienna, July 12-13, 1971, Saudi Arabia, Venezuela, Iran, Iraq, Libya, Kuwait, Qatar, Algeria, Indonesia, and Nigeria decided to rejuvenate this issue. A ministerial committee of representatives of Iran, Iraq, Kuwait, Saudi Arabia, and Libya was formed to develop the bases for participation and to submit recommendations to an Extraordinary Meeting of OPEC members which was to convene on September 22, 1971.[56] A working group was also directed to study a number of legal forms of participation, to recommend methods for assessing costs of the equity share to be acquired by the oil states, to recommend methods for marketing and pricing the governments' share of crude so as to avoid erosion of prices on the international market, and to propose a program for a gradual increase in the governments' equity share in existing concessions over and above the initial prescribed minimum. The prescribed minimum was originally believed to be in the neighborhood of 20 percent.[57] This group was to meet on August 3 to prepare its recommendations.

The manner of participation in the sellers' market was markedly different from that in the buyers' market. Under market conditions favoring the buyers of crude oil, some exporting countries perceived participation in downstream operations as the only realistic means of stabilizing and possibly increasing their revenues. In the sellers' market, however, especially following the Tripoli and Teheran agreements, which resulted not only in raising oil revenues but also in making further revenue enhancements highly attainable objectives, the appetite of the oil states for downstream investments definitely diminished. Now they were focusing on negotiating participation in upstream operations. A noteworthy aspect of the Vienna conference was the Libyan intimation of its "impatience" with what it considered to be the fairly low-keyed negotiating posture adopted by the Gulf states. It was virtually certain that, after concluding agreements of unprecedented dimensions (Tripoli I and II), primarily through riding roughshod over the oil companies, Libya was not about to let the moderates handle the oil industry with kid gloves and demand less than what Libya could get out of them.

As the scheduled conference of September 22 approached, the battlelines of negotiation were being drawn on both sides. When OPEC originally issued its demands for participation, the oil companies dismissed them by saying that the producing countries had not yet defined what they meant by participation. When OPEC developed the framework of demands on this subject, however, the oil companies took the position that participation negotiations were a violation of the Tripoli and the Teheran agreements.[58] The oil states rejected this argument, stating that there was no conflict whatsoever between the participation demand and the obligations of the oil states under the Tripoli and the Teheran agreements.

At an Extraordinary Meeting of OPEC in Beirut on September 22, 1971, the oil states officially unveiled the blueprint of their demands on participation. The member states were given carte blanche to undertake negotiations with the oil companies individually or in groups and were to report their progress, for the purpose of coordination, at the next scheduled conference on December 7 in Abu Dhabi. It was anticipated that member states would adopt the "regionalization" strategy, which had proved so effective during the Teheran and Tripoli II negotiations. The Gulf Group (Saudi Arabia, Iran, Iraq, Kuwait, Abu Dhabi, and Qatar) was expected to carry the brunt of these negotiations on a relatively moderate platform, while Libya was expected to go it alone with a distinctly more hawkish stance. Depsite this hawk/dove categorization, the ultimate goal of all the participant states was to acquire a majority holding of 51 percent in the major concessions. The *modus operandi* of the Gulf states was to acquire initially a minimum participation of 20 percent, which, after some period of stability, was to be subject to upward progression. Libya, following the Algerian pattern, insisted on immediately acquiring 51 percent participation. Nigeria, preferring to conduct its own negotiations in this matter, favored an initial acquisition of 35 percent participation. The remaining three OPEC members were not directly involved in participation. Algeria had already nationalized 51 percent of the operation and production of Compagnie Française de Petroles (CFP). Venezuela, the majority of whose foreign concessions were to expire in 1983, was in the unique position of receiving 80 percent of foreign concessionaires' profits through taxes and royalties. Indonesia, under production-sharing contracts with all its foreign operators, was already receiving 60 to 67.5 percent of crude output and was not interested in immediately altering this arrangement. It should be noted, however, that both the oil states that did not take part in negotiations because of their unique relationship with their foreign concessionaires and those that decided to conduct their own negotiations affected these parlays significantly. First, the Algerian, Indonesian, and Venezuelan agreements with their respective concessionaires, because they were far more lucrative than those prevailing in the Gulf region, were used as justification for the participation of the Gulf oil states in the upstream operations. Second, the deals concluded independently of the Gulf states, especially by Libya, hampered the implementation of the participation agreements in their original form in the Gulf region.[59]

The main rationale for OPEC's participation at the Beirut conference was the legal principle of changing circumstances (*rebus sic santibus*) and the right of sovereign states to act in the best interests of their peoples. On the specifics of participation, a majority of oil states were seeking working interests in the concessions. An acceptance of this arrangement by the oil industry would have enabled the oil states to operate the concessions as their partners. In their altered form, the new demands of the oil states were a radical departure from their original demand of acquiring equity shares in the concessionaire companies themselves. The blueprint also contained two crucial and intractable issues on whose resolution rested the fate of the participation agreements. The first was compensation. The conferees agreed that the net book value of investment should be used to calculate compensation to the companies for the purchase of working interests. In addition, the oil states wanted to make cash payments over a period of several years. The second troublesome issue was the method of marketing and pricing the host governments' share of crude. The oil states wanted adequate safeguards against destablization of crude prices by the influx of "wild" crude in the world market. To minimize this possibility, OPEC not only wanted close cooperation between national oil companies and their foreign concessionaires but also insisted that, at least initially, the bulk of the governments' share of crude move through the integrated channels of foreign concessionaires. Closely intertwined with the marketing of participation crude was the buy-back price, that is, the price which the foreign oil companies were expected to pay the oil states for the latter's share of production. In the participation negotiations, OPEC was expected to press for a "halfway price formula" under which buy-back prices would be set halfway between tax-paid cost and posted prices.

From the viewpoint of the oil industry, these demands were outrageous, although they were a clear reflection of the arrogance formerly exercised by the oil companies when they had the upper hand. The oil industry was not, of course, expected to cooperate fully in expediting the participation negotiations. The industry decided to present a united front. After obtaining an antitrust waiver from the U.S. Department of Justice, industry representatives met in

London to develop their own negotiating position. An important aspect of these cooperative endeavors was creation of a backup supply arrangement to minimize the vulnerability of the independents in case of a curtailment of supplies by Libya. As mentioned in the previous chapter, however, these endeavors did not go very far partly because of the participants' lack of will to stand by them.

Two important incidents should be mentioned whose simultaneous and cumulative impact boosted the Gulf states' bargaining position while heightening the element of uncertainty and thereby weakening the industry's bargaining stance. The first occurred when Italy's state oil company, Ente Nazionali Idrocarburi (ENI), under heavy pressure from Libya, decided to formally initiate participation negotiations with that government. This was a serious setback for the oil industry, which had been especially prepared to deal with Libyan maneuvers against vulnerable oil companies (and ENI was indeed very vulnerable to government pressure).[60] The second incident also involved Libya. On December 7, that country nationalized the holdings of British Petroleum on the pretext that an alleged conspiracy between Britain and Iran lay behind the Iranian seizure of three islands in the Strait of Hormuz: Abu Musa, Tumb, and Lesser Tumb.

No meaningful progress was made on the participation issue until the conclusion of the Geneva I agreement, January 21-22, 1972. OPEC appointed Zaki Yamani, the originator of the concept of participation, as chairman of the negotiating team representing the Gulf states. This team was to negotiate with the oil companies either individually or collectively. The Geneva meeting appeared to conclude in favor of the Gulf states, in that the oil industry did not outright reject the principle of participation. It was clear, however, that in the forthcoming negotiations the issue of compensation would be the most difficult to resolve. The oil companies were represented at this meeting by twelve negotiators from as many companies. Their previous willingness to cooperate was replaced by feelings of secrecy and mutual distrust. So pervasive was this distrust that the company representatives were reluctant to discuss their financial plans and positions in the presence of either their competitors in the industry or the OPEC representatives.[61]

Yamani was asked to make a progress report on his talks with the oil companies at a meeting of OPEC members at the end of February, when the oil states would decide on their course of action. Even before the participation negotiations, it was apparent that Yamani would focus on the American companies. In an interview for the *London Evening Star*, he stated that the French and British companies had resigned themselves to the Arab participation, but the "stubbornness" of the American companies remained the chief stumbling block.

Aramco bore the brunt of these tough negotiations. On February 15, that company made Saudi Arabia its first offer, which provided for 50 percent government participation with Aramco shareholders in developing and operating certain proved but undeveloped oil fields in Saudi Arabia. The offer was promptly rejected. It clearly violated the spirit of participation as envisaged by the Gulf states. Moreover, by its very nature this offer was enticing for Saudi Arabia and Iran, where a sizable portion of oil fields were proved but unexplored, but offered little incentive for other Gulf states. Dissatisfied with the slow pace of negotiations, King Faisal intervened at one point. In a message to Aramco, he warned of unilateral action should the Aramco partners fail to make satisfactory progress toward conclusion of participation agreements.

In view of a continuing impasse, Yamani called for an Extraordinary Meeting of the OPEC Ministerial Conference on March 10. The oil states let it be known that they were about to discuss drastic actions to enforce participation. At this point Aramco caved in and accepted the principle of 20 percent participation in its concession. With this clearcut and perhaps most significant victory under their belt, the Gulf states faced another formidable task, that of bridging the gulf on the specifics of participation. For instance, the two sides held drastically divergent views on the price to be paid for the host governments' 20 percent share. The Gulf states wanted to compensate the oil companies on the basis of net book value of above-ground assets, while the oil companies felt their compensation should include proved oil reserves as well as discounted future profits from the sale of those reserves. Yamani was to conduct further negotiations on behalf of the six Gulf states on matters of detail.

Aramco's acceptance of 20 percent participation was a tribute to the growing (and understandably begrudging) adaptability of the oil industry under the mounting uncertainties of the sellers' market. The significance the Gulf states attached to the issue of participation and the extent to which they were willing to go to attain it were evidenced by the fact that they ostensibly planned to implement it by force if necessary.[62]

Yamani held a series of negotiations with the oil companies in San Francisco, Riyadh, Geneva, and London, but the two sides continued to be a world apart on the cost of acquisition by the host governments, the buy-back price, and the timing for escalating the Gulf states' share of participation to 51 percent. During the initial stages of these negotiations, there was an impressive display of cohesion among the Gulf states. Toward the middle of 1972, however, two incidents took place, one of which had the potential to improve the oil companies' bargaining position. On June 24, the Shah announced that Iran was dropping out of the participation negotiations. This announcement initially caused a certain amount of dismay among the Gulf states, but the Iranian officials assured their counterparts at the Twenty-Ninth Conference of OPEC in Vienna, June 26-27, that the Iranian withdrawal was not directed against OPEC's participation negotiations but was a parallel move taken in light of the special relationship that prevailed between Iran and its consortium. The second incident was Iraqi nationalization of Iraq Petroleum Company (IPC) on June 1, a move that further boosted the position of the Gulf states. In the past, OPEC had stayed out of this dispute, labeling it primarily a domestic matter of Iraq. This time, however, the issue of contention between IPC and Iraq directly involved OPEC. The government of Iraq, which nationalized the North Rumaila oil field and IPC's other unexploited concession acreage under Law No. 80 of 1961, wanted to compensate IPC on the basis of net book value of work undertaken. IPC insisted on receiving the equivalent of 7 percent of all future oil production by the Iraq National Oil Company (INOC) as compensation for its loss of rights concerning oil discoveries made before 1961. As the preceding discussion makes clear, this was one of the major issues of contention in the participation negotiations. OPEC labeled Iraq's action "a lawful act of sovereignty to safeguard its legitimate interests."[63] The quid pro quo effect of Iraqi action and OPEC's endorsement of it were clearly understood by the oil industry. It was a clearcut signal of what the other Gulf states might do if no agreement was reached on the participation question.

A comprehensive agreement between OPEC and the oil industry was reached in New York on October 5, 1972. The "General Agreement on Participation," as it was called, included the following provisions:

(1) It allowed for an initial 25 percent participation by the governments concerned immediately after finalization of the relevant agreements. This was to be increased to 30 percent on January 1, 1979; to 35 percent on January 1, 1980; to 40 percent on January 1, 1981; to 45 percent on January 1, 1982; and to 51 percent on January 1, 1983. It was to remain at 51 percent until the expiration of the concessions. As previously noted, the Gulf states' original demand was for 20 percent initial participation. During the final phase of negotiations, when the timetable for 51 percent government takeover was under discussion, Yamani hoped to gain majority control by 1979, but the oil companies preferred a longer transition period. As a tradeoff, the parties agreed to extend the period of transition in exchange for a boost from 20 to 25 percent in the initial share of participation.

(2) The Gulf states agreed to pay for their equity share on the basis of "updated net book value," which was interpreted as "written down book value adjusted for inflation." The oil companies were not to be compensated for underground oil reserves or loss of future profits.

(3) In order to facilitate the transfer of ownership with minimal confusion in the international oil market and with least hardship for all parties, the oil companies agreed to purchase "bridging crude" at close to the prevailing market price and "phase-in crude" at slightly less than the market price.[64]

This agreement had to be signed and ratified by at least three states, and the signatories had to negotiate separate agreements with their respective concessionaires in order to implement it. A major snag was encountered when, on

October 9, Libya announced its conclusion of a 50/50 joint-venture with ENI. Libya circulated the text of this agreement at a meeting in Kuwait, where the Gulf oil ministers had gathered to study the details of the participation agreement concluded by Yamani.

At a different time, an agreement of this magnitude would have provided the stability long sought by the oil industry, and the oil-exporting states might also have been content with the qualitative changes brought by the participation agreements. But the growing uncertainties of the sellers' market never allowed implementation of these agreements in their orginal form. Of the five Gulf states—Saudi Arabia, Kuwait, Iraq, Qatar, and Abu Dhabi—four initially signed the agreement. Iraq objected to the updated book value formula since its acceptance of this mechanism would have required it to compensate IPC for the June 1 nationalization on a similar basis. Kuwait failed to ratify it when the Kuwaiti National Assembly rejected it, and that country decided to press for a majority share long before the scheduled date. Libya also continued its single-handed efforts to acquire a majority share. This time Libya used Bunker-Hunt as a guinea pig for an immediate 50 percent share. In view of Hunt's continued resistance, Libya nationalized that company in June 1973. In August, Occidental acquiesced to the Libyan takeover of a 51 percent share. Libya followed with a 51 percent nationalization of the Oasis group, which was comprised of Marathon, Continental, and Amerada Hess. The terms of the Libyan takeover were bound to create a stampede among the Gulf states for similar deals. Specifically, these terms included an immediate 51 percent participation, as opposed to the phased participation negotiated by Yamani; compensation for Oasis based on net book value, as opposed to the updated book value agreed to for the Gulf states; and a buy-back price of $4.90/b for 51 percent of Occidental's production, as opposed to the average buy-back price of $2.50/b for 25 percent production for Gulf states.[65] Nigeria negotiated a 35 percent interest in June, and Kuwait was negotiating a deal which would top the Libyan share. In the aftermath of the Arab/Israeli war of October 1973, Iraq nationalized the holdings of Royal Dutch, Exxon, and Mobil in Basrah Petroleum Company. Iran in the summer of 1973 concluded an agreement with the Iranian Consortium, which was approved by the Majlis (parliament) in July of that year. This agreement amounted to 100 percent takeover, and the foreign concessionaires were retained to run the consortium.

By November of 1973, OPEC declared the schedules of 51 percent participation negotiated by Yamani "insufficient and unsatisfactory." Under the acute uncertainties of the sellers' market in 1974, Saudi Arabia, Kuwait, Qatar, and Abu Dhabi escalated their participation level to 60 percent. In December 1975, Kuwait negotiated a 100 percent takeover of Kuwait Oil Company (KOC). In the years 1975 through 1977, Qatar and Saudi Arabia also concluded similar arrangements with their concessionaires.

The period dealt with here was marked by a series of events which were unusual even by the standards of oil affairs of the recent past. The coincidental and almost simultaneous occurrence of the Arab oil embargo imposed by OAPEC and the unilateral OPEC price increases of October and December 1973 created the appearance of an energy crisis in the Western industrial societies. Moreover, OPEC decisively emerged as the advantaged actor, frequently establishing its rationale for renegotiating, at will, agreements which it had recently concluded. Despite recurrent in-group conflicts and squabbles between the price moderates, led by Saudi Arabia, and the price hawks, led interchangeably by Iran and Libya in conjunction with Iraq and Algeria, OPEC's solidarity on intermittent increases in the prices of crudes largely remained undented. Even though their rationale for price escalations was universally deplored, especially by the Western industrial consuming nations, OPEC members sincerely believed that oil had been an underpriced commodity until then, that their peoples' standards of living were rising at a much slower pace than those of the industrial countries, and that they were doomed to remain underdeveloped unless something was done before oil reserves ran out or before alternative sources of energy obviated the need for oil.[66]

Two remarkable phenomena of this period were the chaos emanating from the oil embargo and the quantum price escalations of 1973 together with the responses of the industrialized consumers, some of which were contradictory in nature. The chaos resulted largely from the unprecedented nature of the shocks of 1973 and the total unpreparedness of the international economic system. Precisely because the Western Europeans and Japan did not anticipate these actions, their immediate response was unilateral actions to minimize the grave economic consequences of their inordinate dependence on OPEC oil. Their political objectives of developing medium- and long-range collective policies to reduce their energy dependence were assigned secondary significance. It was not until 1974 that cooperative arrangements between industrial consumers were negotiated after much in-group wrangling. Even then, the Western Europeans and the Japanese remained nervous about generating antagonistic policies on the part of OPEC. The United States, primarily because its economic dependence on OPEC oil was minimal compared to that of its allies, doggedly pursued its concomitant political objectives of dismantling OPEC and establishing cooperative arrangements among the industrial consumers. Partly because of the American confrontational posture toward OPEC, however, the allied response to cooperative policies was so slow to come.

Another significant phenomenon of this period was that the economic ability of the oil states to manipulate prices of crude and, to a certain extent, to adjust production rates was soon translated into political power. For the Arab states, imposition of the oil embargo was a political action whose economic consequences benefitted them. OPEC used its economic ability to escalate prices to try to attain the political objective of creating the so-called New International Economic Order. Thus, in the aftermath of 1973, the intermingling of political and economic variables transformed the war of nerves between the oil states and the Western industrial states led by the United States, the guardian of the political and economic status quo of the non-Communist world. At one time, roughly between 1971 and 1973, this war of nerves may have been over the prices of crude oil. But between 1974 and 1975, this tussle was transformed into a struggle over how much of the political and economic status quo the United States was willing to alter. Perhaps the abortive attempts to create NIEO (through the so-called North-South dialogue) were the ultimate symbolization of that struggle. In the meantime, OPEC and the Western industrial countries continued to use these meetings for rhetorical one-upmanship by blaming each other for prevailing economic chaos and doing virtually nothing to ameliorate it.

The last quarter of 1974 and the first quarter of 1975 were testing times for OPEC. This period was marked by a slackening of demands and by an increase in supplies. Moreover, largely because of implementation of various shades of participation agreements in the Gulf region, the national oil companies had, for the most part, taken over upstream operations, with the foreign concessionaires primarily operating as producers under contract. This situation perfectly suited the conceptual scenario repeatedly depicted by the noted oil economist M.A. Adelman for the collapse of OPEC as a cartel. According to Adelman's scenario, a system of high competition would prevail if the oil states were to assume the role of sellers of their commodity and the multinational corporations were to serve as "producers under contract and as buyers of crude to transport, refine and sell as products." The price then would be only "bare cost" and not the "tax-plus-cost." The oil states would be forced to choose between setting production quotas and selling their oil at discount prices, thus triggering similar responses from other producers. Such unilateral action, in an environment of competition and mutual distrust, noted Adelman, would result in the crumbling of the OPEC cartel.[67] A theme similar to Adelman's was echoed by Frank Gardner of *Oil and Gas Journal* when he wrote in 1975 that if the nationalization of the oil industry continued, at a point in the near future twelve national oil companies of the OPEC states would be competing for their respective shares of the world oil market. And with the dwindling demand and surging surpluses which then prevailed as a result of the great recession, Gardner believed OPEC would find it increasingly difficult to cooperate either on price controls or on production quotas. This, in Gardner's view, would likely result in the breaking up of OPEC.[68] The predictions of Adelman's scenario and Gardner's analysis did not materialize, however, because a number of major OPEC producers were able to absorb the demand shortfall among themselves. In 1974-1975, for instance, Saudi Arabia reduced its production by 40.5 percent, Venezuela by

18.3 percent, Iran by 19.4 percent, and Nigeria and Kuwait by 13.4 percent.[69] Both Gardner's and Adelman's arguments also ignore the fact that in the sellers' market the oil states were not likely to encounter a situation in which they would be forced to shave profits to increase sales. In reality, whenever the sellers' market began to show signs of over supply, the producers tinkered with their levels of production in order to sustain their levels of profit.

The foregoing analysis would not be complete without a brief mention of arguments presented by James Akins, former director of the State Department's Office of Fuels and Energy and later the U.S. ambassador to Saudi Arabia, who was in major disagreement with Adelman and Gardner and whose analysis of OPEC's behavior in this period seems to have been more in line with prevailing reality. Akins stated that, contrary to arguments presented by a few academicians that OPEC was a producer cartel that would be broken up by greed from within the organization, the oil states were examining the proposition that they might be making money too fast. According to Akins, OPEC did not have forces competing for a higher share of the market and was far from unstable; and anyone who assumed that the oil producers would engage in cutting each other's throats for an increased share of the market "shows impressive ignorance of contemporary policies and even a lack of knowledge of current events." Akins also observed that the petroleum-exporting nations were interested in keying oil production so as to produce income at a rate they could spend profitably and wisely, and were increasingly striving to raise income per barrel as opposed to escalating levels of production. The rationale underlying such careful oil policies was the belief by the exporting countries that oil in the ground would appreciate at least as fast as bank deposits and much faster than gold.[70] Amplifying on this theme elsewhere, Akins wrote that scenarios of OPEC's demise are read not only in the West but also in OPEC countries, and that such analyses only increase the already firm resolve of the oil states to avoid such a development. The oil states might be interested in maximizing their profits, but they were aware that this could be done most easily by raising prices. No OPEC country, according to Akins, was interested in "breaking" world oil prices, thereby breaking OPEC itself.[71]

The oil states remained fully cognizant of the pernicious impact of competition among themselves. Conservation measures, as adopted by Venezuela, Kuwait, and Libya, especially during the last quarter of 1974 and the first quarter of 1975, unwittingly became a tool for minimizing the dysfunctional effects of intra-OPEC competition. In addition, the oil states continued their quest for further safeguards against the possible collapse of OPEC. At the Ninth Arab Congress of 1975, preventive measures were very much on the minds of officials from the oil states. The oil minister of Abu Dhabi, Mana al-Otaiba, urged OPEC members to agree on a method for prorationing oil production as a solution to oversupply. Their inability to do so, said he, might force them either to lower prices unilaterally or to bring about production cutbacks. The most significant topic of this congress was a series of proposals submitted by Kuwait with a view to further consolidating OPEC. Chief among these were: (1) methods to remedy the harmful effects of the fall in the value of the dollar; (2) proposals to set up a collective system of protective financial arrangements to assist any OPEC member state exposed to outside economic pressures because of its implementation of the organization's policies and decisions; (3) proposals for greater cooperation, coordination, and mutual support between the national oil companies of the member states; and (4) a study of means of cooperation and coordination with a view to maintaining the balance between supply and demand in the world oil market to prevent an oil surplus that might depress the price levels.[72]

The tenacious Adelman in September 1975 came up with another scheme which, he felt, would eventually bring about the demise of OPEC. He suggested that the U.S. government could tempt the oil exporters to cheat on their high fixed prices by making cheating easier. This could be done by setting up a system of secret, competitive bidding for licenses to import oil into the United States. Under this secret system, he predicted, that first one member of the cartel and then another would effectively cut prices in a bid for a larger share of the U.S. market. Adelman was confident that such events, once started, would be self-propelling. He acknowledged that getting the price cuts started—breaking the discipline of the cartel—was the least certain part of his strategy.[73] Again, there are inherent weaknesses in Adelman's scheme. First, he assumed that the oil states were yearning for additional oil markets. The prevailing reality did not support this assumption. Second, as previously noted, during this period the oil states were increasingly turning toward conservation and were interested in raising their income *per barrel* through price escalations rather than in increasing the rates of production. By applying any rule of economics, it is sheer folly to expect that in the sellers' market the oil states would raise their production levels to gain access to the U.S. market or that they would indulge themselves in covert price cutting, as suggested by Adelman.

The thrust of the foregoing analysis is that OPEC disheartened its detractors and ill-wishers by outliving a variety of scenarios of its possible collapse and other schemes, such as the American-initiated abortive attempts to establish an organization of petroleum-importing countries emulating OPEC, whose purpose would be to minimize OPEC's ability to manipulate the oil market.

The price hikes of October and December 1973 were the peak performance of OPEC until 1979. This organization was able to introduce only token increments in royalties and tax rates during 1974 and 1975 while the price of OPEC oil remained frozen. Considering that during these two years the inflationary rates continued to move upward, the real price of oil during 1974 and 1975 may have experienced a slight decline. In addition, the inability of the oil states to insulate their economies from international inflation also meant that they were to suffer from the worst effects of their price decision between 1976 and 1978.

5. The Oil Embargo

The imposition of the oil embargo in October 1973 by the member states of the Organization of Arab Petroleum Exporting Countries (OAPEC)[1] underscored three painful realities for the Western industrial countries. First, the embargo was the most concrete evidence of the ability of the Arab oil states to flex their muscles toward attainment of their long-cherished political goal of resolving the Arab/Israeli conflict. Second, the embargo demonstrated that the energy vulnerability of the Western European countries and Japan could be used to create considerable dissension within the Western economic alliance, with deleterious spillover effects that might weaken the Western military alliance. Third, the oil embargo convinced policymakers in the industrial countries that they could no longer operate on a *laissez-faire* basis on energy matters, leaving crucial decisions regarding their oil supplies primarily in the hands of the multinational oil corporations. In response, the industrial countries scrambled to develop concrete domestic energy policies, the focus of which was to promote energy self-sufficiency through the efficient use of energy and to decrease dependence on oil through increased use of such alternative sources as coal and nuclear energy. In addition, these countries eventually agreed upon multilateral cooperative arrangements for emergency oil-sharing and for maintaining domestic oil stocks as a buffer against future oil shortfalls emanating from political turmoil in the Middle East or from embargoes.

The oil embargo, which lasted from October 1973 to March 1974, intensified supply-related uncertainties for the industrial West and forced these countries to shift their policies in a more pro-Arab direction with regard to the Arab/Israeli conflict. The chief significance of the embargo was not that it could be repeated in the future. In fact, under current political and economic realities, the likelihood of its repetition is virtually nil. But the oil embargo continues to be an uneasy reminder of the failure of the industrial countries to find dependable avenues of energy supply. Although these countries are much better prepared to deal with a crude shortfall in the mid-1980s than they were in the early 1970s, their dependence on foreign sources of supply which lie beyond their realm of control and which could be eliminated through concerted political action or upheaval remains a frightening reality.

The success of the oil embargo in bringing an end to the Middle Eastern stalemate,[2] i.e., a no-war-no-peace situation, and in creating new momentum toward a political settlement of the conflict was based on several realities of the world oil market. First, the increased effectiveness of OPEC between 1971 and 1973 drastically tilted the balance of power in favor of the oil states, and reduced the status of the oil companies almost to that of buyers and sellers of oil. The Arab oil states served their apprenticeship within the framework of OPEC and were ready to utilize the "OPEC experience" to resolve the Arab/Israeli conflict without jeopardizing their oil revenues in a highly buoyant market. Second, U.S. oil imports from the Middle East were on the rise, a pattern expected to continue in the foreseeable future. In the absence of abundantly available non-OAPEC supplies to replace the potential loss of Arab oil, the United States remained significantly vulnerable to an oil embargo. Finally, the inability of the Western industrial countries to establish a consumer organization to counter OPEC, or at least to make OPEC's price escalations between 1971 and 1973 more difficult, fully exposed the impotence of the Western alliance. This impotence may have convinced the Arab oil states that the West would similarly fail to agree on a retaliatory response even to such a drastic action as an embargo.

Background

On two occasions prior to 1973, unsuccessful attempts were made to use Middle Eastern oil on a limited scale as a political weapon. The first such attempt was in the aftermath of the 1956 joint invasion of Egypt by England, France, and Israel. At that time the flow of oil was disrupted because of the Suez Canal closure and the blowing up of one of the pipelines from Iraq to Lebanon through Syria. About two-thirds of the Middle Eastern exports to Europe had to be rerouted. The results were minor inconveniences in the form of temporary shortages and moderate price increases. The second time the Arab states tried to use oil as a political weapon was in the wake of the outbreak of war between Egypt and Israel in 1967. Crude oil production in Saudi Arabia and Libya was brought to a halt by strikes by oil workers. Iraq and Kuwait, on the other hand, suspended production as a matter of official policy, a suspension later replaced by a selective embargo on oil exports to the United States, Britain, and West Germany. There are several reasons why disruption of supplies in 1956 and 1967 remained so insignificant. First, it should be remembered that in the 1950s and 1960s the oil market remained a buyers' market. The overall patterns of Western European and Japanese dependence on Arab oil on both occasions were by no means insignificant. In 1956 and 1967, 65 and 68 percent of Western European energy supplies, respectively, were met by Arab oil; Japanese imports of Arab oil remained at 56 percent in 1956 and 50 percent in 1967. But, in the prevalence of the buyers' market, even the substantial dependence of the West did not appear to strengthen the Arab oil states' resolve to make their oil weapon work. The second reason the disruption of supplies was not significant was that the United States was not heavily dependent on Arab oil. In 1956, the U.S. filled 23 percent of its energy needs from the Arab countries, but by 1967 this dependence had been reduced to 8 percent, largely because of import quotas imposed by the Eisenhower Administration.[3] What seriously undermined all efforts to use the oil weapon in 1956 and 1967, however, was U.S. determination to break any attempted embargo in tandem with American willingness to make its own oil available to other embargoed countries. Finally, the remarkably sophisticated capability of the oil companies to distribute to crude-short countries nonembargoed oil, and even embargoed oil that was available because of "leaks," did its share to nullify any potentially harmful effects from supply disruptions on both occasions. The Arab oil states, especially in 1967, encountered considerable losses in production and revenue, the chief beneficiaries of which were the non-Arab producers.[4]

By 1973 the political realities of the Middle East and the economic realities of the world oil market presented a markedly favorable picture for the Arab oil states should they decide to use their oil weapon to influence the behavior of the industrial consuming nations. The traditional vestiges of colonialism and imperialism had disappeared and the newly independent states were involved in nation-building and its attendant problems. Politically speaking, the conflict between the Nasserites and the monarchists, a predominantly divisive force in the 1950s and 1960s, had lost its vigor by the 1970s, with the death of Nasser. The Arab oil states were also successful in substantially diminishing the traditionally powerful status of the Western oil companies as a result of their maneuverings on the price issue and, more importantly, as a result of the participation agreements of 1973. Israel, however, the only symbol of Western "injustice" whose status remained totally unaffected by the newly-acquired oil power of the Arabs, continued to occupy their land and seemed to be bent upon changing its fortunes of war into permanent borders by using the excuse of acquiring "secure" and "defensible" borders.[5]

With the powers of the international oil companies diminished, with the demise of Nasser, with the minimization of traditional conflict between the Nasserites and the monarchists, and with the obsolescence of the conspiratorial strategies allegedly associated with the Nasser regime, the only significant and unresolved conflict that remained in the Middle East was the one between the Arabs and the state of Israel, namely, Israeli withdrawal from Arab lands and resolution of the Palestinian question.

A significant spillover effect of the growing power and prestige of OPEC was the rejuvenation of Arab demands for the use of oil as a political weapon. Frequent sources of such clamor were mostly Arab states that had little or no oil to commit to this exercise or those that had the oil but no real intention of using it on behalf of Arab causes.[6] The focal point of this demand was OAPEC, whose members had acquired considerable experience in negotiating with the oil companies both within and outside the framework of OPEC. Those who pressed for the use of oil as a political weapon hoped that the newly-acquired economic power of oil might bring about a political solution to the Arab/Israeli conflict that would satisfy the Arabs.

Ironically, an original, explicit purpose of OAPEC was to keep the oil question out of politics.[7] Saudi Arabia remained the single most ardent spokesman for this point of view, while Algeria, Iraq, and (after 1969) Libya advocated equally vehemently the use of oil as a political weapon for the overall resolution of the Arab/Israeli conflict. The insistence of the Saudis and other oil monarchies on separating oil from politics was based on their view that any application of the Arab oil weapon in the ample availability of non-Arab oil was akin to committing economic suicide. The abortive selective embargo of 1967 was a glaring example of how oil consumers could easily switch to other oil sources. The reluctance of the monarchies to use the oil weapon had a vital political dimension as well. Prior to the 1970s there seems to have existed a symbiotic relationship between the conservative Arab states, such as Saudi Arabia, Kuwait, Iran, and other Persian Gulf states, and the United States (and earlier with the United Kingdom as well). Essentially, this relationship seems to have been based upon a tacit understanding that the United States would guarantee the survival of the conservative regimes, while the latter would serve as secure channels for the oil needs of the United States and its industrial allies. Within this relationship, the pendulum of advantage swung in favor of the United States, since it was the monarchical regimes whose survival was at stake (or so they themselves perceived) during the 1950s and 1960s, a period when Nasserite forces were allegedly plotting the overthrow of all pro-Western regimes in the Middle East. This asymmetrical association not only prevented the Arab monarchies from trying to sway what they perceived as an inordinately pro-Israeli American foreign policy in their favor but also kept them from seriously considering the use of oil as a political weapon against the West.

The 1970s, however, removed the specter of political instability from the Arab monarchies. Their treasuries were recording enormous oil surpluses. At an opportune moment in the near future, the Arab oil states of all political leanings would be ready to trigger a psychological war by imposing an oil embargo on Israel's arch-supporter, the United States, and by reducing crude oil supplies to those countries whose traditional policy was either less favorable or neutral toward resolution of the Arab/Israeli conflict.

Imposing the Embargo

Throughout the late spring and summer of 1973, the clamor for use of the "oil weapon" was increasing in the Arab world. Economically speaking, its potential use was in no way expected to hurt the Arab oil states. If anything, a heightened crude shortfall, by all calculations, would be likely to enhance per-barrel income. The dependence of the Western industrial countries on Arab oil reached a new peak in 1973. Western Europe imported 45 percent of its oil needs from the Arab states; Japan, 33 percent; and the United States, about 5 percent.[8] Oil revenues of all Arab states were on the rise. As shown in Table 13, with the exception of Iraq and Libya, all oil producers recorded substantial revenue gains from 1971 to 1972, with Algeria and Saudi Arabia recording the largest gains. The loss of revenue by Iraq was due to a production shutdown when the government nationalized the northern oil fields of Iraq Petroleum Company; the loss of Libyan revenue was due to production cutbacks imposed by the government.

In view of these facts, the conservative oil states had no overriding reason for not using the oil weapon for political purposes. In April, while visiting Washington, Saudi Oil Minister Zaki Yamani officially expressed his government's willingness to use oil to attain political objectives. He intimated to American officials that his country would find it difficult to increase oil production if the United States failed to use its influence with Israel to extract a political solution satisfactory to the Arab States.[9] This action was followed by expression of dissatisfaction with U.S. Middle East policy by Saudi Arabia's King Faisal, who used private as well as public channels of communication to convey his feelings to the U.S. Government.[10]

Accustomed to the Arab penchant for rhetoric, which almost invariably was not followed by substantive actions, the United States chose to respond to statements by Saudi officials by sustaining a policy of inaction. At the same time, in the Middle East the resolve of the Arab oil states to proceed with an embargo, if necessary, became more apparent at a meeting held on September 4. This meeting was crucial because it was here that the specifics of division of labor among various Arab states were fully explored. The hardliners, such as Algeria, Libya, and Iraq, felt that full or partial nationalization of the assets of American oil companies would be the best retaliation for the pro-Israeli policies of the United States. The moderates, such as Kuwait and Abu Dhabi, expressed their willingness to employ oil as a political weapon under two conditions: first, that such action should involve all oil states; and second, that the "front-line" Arab states, such as Egypt and Syria, demonstrate their willingness to fight for recovery of their occupied land. If these conditions were not fulfilled, the moderate states preferred continued application of their oil as a "positive

Table 13. Oil States' Exports and Oil Payments, 1971 and 1972

Producer	Exports (million b)			Oil payments ($ million)			Oil payments (¢/b)		
	1971	1972	% change	1971	1972	% change	1971	1972	% Change
Mideast									
Kuwait	1,168.7	1,175.7	+0.6	1,399.8	1,656.8	+18.4	119.7	140.9	+17.7
Saudia Arabia	1,707.3	2,162.7	+26.7	2,148.9	3,106.9	+44.6	125.9	143.7	+14.1
Iran	1,562.2	1,751.9	+12.1	1,944.2	2,379.8	+22.4	124.6	135.8	+9.0
Iraq	593.5	381.6	−35.7	840.0	575.0	−31.6	141.5	150.7	+6.5
Abu Dhabi	338.5	384.2	+13.5	430.7	550.9	+27.9	127.2	143.4	+12.7
Qatar	156.5	176.3	+12.7	197.8	254.8	+28.8	126.4	144.5	+14.3
Others	179.0	184.6	+12.7	192.6	222.6	+15.6	110.1	120.6	+9.5
Total	5,705.7	6,217.0	9.0	7,154.0	8,746.8	+22.3			
Other									
Libya	988.9	812.9	−17.8	1,766.0	1,598.0	−9.5	178.6	196.6	+10.1
Algeria	276.0	373.0	+35.1	350.0	700.0	+100.0	126.8	187.7	+48.0
Nigeria	531.4	628.0	+18.2	915.0	1,174.4	+28.3	—	—	—
Venezuela	1,205.9	1,132.8	−6.1	1,702.0	1,948.0	+14.5	141.1	171.9	+21.8
Total	8,707.9	9,163.7	+5.2	11,887.0	14,167.2	+19.2			

Source: *Oil and Gas Journal*, Oct. 29, 1973, p. 55.

weapon," i.e., as a generator of capital for the front-line Arab states.[11]

On October 6, heavy fighting erupted between Arab and Israeli forces along the Suez Canal and Golan Heights ceasefire lines. By starting the war, Egypt and Syria had unequivocally demonstrated their will to fight to bring about an Israeli withdrawal from the occupied territories. Now is was up to the moderate Arab oil states to do their part. The decision to proceed with an embargo, though of enormous magnitude, was not half as intricate as the task of working out its specifics. To begin with, under the highly charged environment at the outbreak of hostilities, the task of translating the Arab consensus to use oil as a political weapon was problematic. There was always the danger that the specifics of embargo preferred by the moderate states might be dismissed by the hardliners as a "lukewarm" commitment to the Arab cause. At the same time, carrying the embargo decision too far by totally cutting off oil production might have forced the West to resort to military action against the Arabs to secure supplies. The realities of international politics and economics made it essential that political extremism or emotive pan-Arabism not be allowed to take charge of specific policies governing the embargo. In imposing the embargo, the Saudi and Kuwaiti preference for pursuing a moderate course prevailed. This moderate approach further complicated the task of dealing with different countries. For instance, in the industrial world, the United States and France adopted entirely different policies toward the Arab/Israeli conflict and thus had to be dealt with differently. Even in the nonindustrial world, such traditional friends of the Arabs as India and Pakistan could not be lumped with the rest of the Third World countries. One alternative for the Arab oil states was to impose a selective embargo against the United States, which would be difficult to enforce and might not have a telling effect. Memories of the fate of such an embargo in 1967, however, were too vivid for it to be considered. Another option was to impose an embargo on all Western countries. This option would have alienated Western Europe and Japan, whose sympathies and support were vital for the Arabs if they were to put political pressure on the United States. Besides, as previously noted, the option of imposing an embargo on all the Western states was militarily dangerous. Ultimately, the OAPEC states decided to adopt a triple-barreled strategy. The first objective was to avoid antagonizing those states that were friendly to them. Second, the strategy was aimed at enticing those countries that were leaning toward Israel or that preferred neutrality to adopt pro-Arab policies concerning the conflict. Third, this strategy was intended to punish those states that were perceived by the Arabs as allies of Israel. Obviously, considering the support it had extended to the Jewish state, the United States was the focal point of Arab anger.

After the October 16 meeting of OPEC in Kuwait, where members announced quantum leaps in the prices of oil (see Chapter 4), the Arab oil states met on October 17 under the auspices of OAPEC and agreed to cut production

Table 14. Estimates of Arab Oil Cutbacks, as of October 26, 1973 (1,000 b/d)

Producer	Sept. Output	Est. Oct. 26 Output	Est. Cutback
Saudi Arabia	8,300	6,500	1,800
Kuwait	3,200	2,340	860
Iraq	2,000	1,500	500
Abu Dhabi	1,400	1,220	180
Qatar	582	462	120
Neutral Zone	582	524	58
Libya	2,300	2,100	200
Algeria	1,050	900	150
Other[a]	1,050	800	250
Total	20,464	16,346	4,118

Source: *Middle East Economic Survey,* Oct. 26, 1973.

[a]Egypt, Syria, Bahrain, Dubai, and Oman.

by a minimum of 5 percent, using the September 1973 production level as a base. The communiqué issued at the conclusion of this meeting promised to introduce similar reductions during the following months until the Israeli withdrawal from Arab territories occupied since 1967 was completed and the legitimate rights of the Palestinians were restored. A system was established whereby each country was to instruct its operating companies of its desired pattern of exports for any given month. The companies, under the threat of penalty for violation, were required to carry out each government's instructions. This communiqué was signed by Saudi Arabia, Kuwait, Libya, Algeria, Qatar, Bahrain, Abu Dhabi, Egypt, and Syria. Iraq refused to sign the communiqué on the grounds that it did not contain harsh enough measures against the United States, such as "complete liquidation" of American economic interests, withdrawal of all funds invested by the Arab states in the American economy, and severance of diplomatic ties between all Arab oil states and the United States. Iraq imposed a total embargo on crude supplies to the United States and Holland, and nationalized the interests of Royal Dutch-Shell in the Iraqi operations of BP.

One day after issuance of this communiqué, the Arab oil states tightened their oil weapon in three ways. First, instead of introducing 5 percent cuts for the first month, some participating countries announced reductions of 10 percent. Second, all participants announced a total embargo on shipments to the United States, and some announced a total embargo on those to the Netherlands. Third, OAPEC members warned that nations caught transshipping oil to the United States would also be embargoed.[12] According to estimates compiled by an authoritative source, toward the end of October the total reduction in Arab oil supplies may have been as high as 4 million b/d, or 20 percent below the level of

Table 15. Chronology of Restrictions Imposed by OAPEC on the United States and the Netherlands, 1973

Date	Producer	Production Cutback (%)	Embargo Against: United States	Embargo Against: Netherlands	Other Action
Oct. 18	Saudi Arabia	10			
	Qatar	10			
	Libya	5			
	Abu Dhabi		X		
	Algeria		X[a]		
Oct. 19	Libya		X		
Oct. 20	Bahrain	5	X		Cancellation of 1971 U.S. Navy base agreement
	Saudi Arabia		X		
	Algeria	10			
Oct. 21	Kuwait	10	X		
	Dubai		X		
	Qatar		X		
	Bahrain		X		
	Algeria			X	
Oct. 22	Iraq				Nationalization of Royal Dutch-Shell interest in BP
Oct. 23	Kuwait			X	
	Abu Dhabi			X	
Oct. 24	Qatar			X	
Oct. 25	Oman		X		
Oct. 30	Libya			X	
	Bahrain			X	
Nov. 2	Saudi Arabia			X	

Source: *U.S. Oil Companies and the Arab Oil Embargo: The International Allocation of Constricted Supplies,* report prepared for the Subcommittee on Multinational Corporations of the Senate Committee on Foreign Relations by the Federal Energy Administration, Office of International Energy Affairs (Washington, D.C.: GPO, 1975), p. 15.

[a]Originally imposed Oct. 6.

output in September, equivalent to a shortfall of some 12 percent of the total volume of oil moving in world trade.[13] A detailed breakdown of the Arab production cutback of October 1973 is provided in Table 14. By early November, cutbacks and other embargo-related actions were firmly in place. Table 15 shows actions taken by the Arab oil states from October 17 through November 2, 1973, a span of sixteen days.

If one considers how difficult it is to impose a partial embargo without losing its potency through major leaks, the Arab oil states established a track record for efficiency. Saudi Arabia, for example, kept detailed information and maintained the "strictest control" over every barrel of oil leaving its ports. Ship captains accepting Saudi oil were required not only to sign affidavits stating their destination but also to report their delivery by cable to Aramco. This procedure was followed to preclude mid-ocean diversions. Moreover, Arab diplomats abroad monitored public records of oil imports by country of origin.[14] Lest there be any doubt concerning the embargoed, semiembargoed, and nonembargoed countries, the Arab oil states also issued the following list of categories, along with the volume of oil allocated to each:

(1) Preferred, or nonembargoed, countries were allowed 100 percent of supplies at September 1973 levels. These included Arab countries such as Egypt, Jordan, Lebanon, and Tunisia; Islamic countries such as Pakistan, Turkey, and Malaysia; and other states such as France, Spain, Britain, India, Brazil, and those African states that broke off diplomatic ties with Israel. In December 1973, the Arab oil states transferred Arab, Islamic, and African importing countries from the preferred to a new category, "most favored nations."

(2) Neutral or semiembargoed nations received a reduced percentage of September 1973 levels. Belgium and Japan were originally listed in this category.

(3) Embargoed countries received no crude oil or refined products. This included the United States, the Netherlands, Portugal, and South Africa. In order to assure a total embargo on these countries, the Arab oil states also imposed an export ban on a variety of countries which were known for their previous roles as intermediate suppliers of U.S. markets: Trinidad, the Bahamas, Dutch Antilles (Curacao), Canada, Puerto Rico, Bahrain (which normally supplied 50,000 b/d to the American navy), Guam, and Singapore (another outlet for U.S. naval supplies). Shipments were also curtailed to certain specific refineries which supplied American markets, including those in Italy, those in Greece (which supplied the 6th Fleet), and one refinery in southern France.[15]

In November, the Arab oil states felt the need for a mechanism to assure that the burden of production cutbacks would be uniformly applied and not excessively borne by just a few states. Upon the initiative of Saudi Arabia, a

November 4-5 OAPEC meeting was called in Kuwait for this purpose.[16] The participants decided to raise the cutbacks to 25 percent immediately. This percentage included volume reductions as a result of embargoes against the United States and Holland. This decision was also aimed at significantly lowering production of light crude in Libya, Algeria, Abu Dhabi, and Qatar, which was in high demand in the West, especially in the United States. Moreover, any reduction in short-haul Mediterranean crude from Algeria and Libya was bound to tighten the oil market further. The cumulative result of this decision was an estimated decrease of 28.5 percent below the September level. In a radical departure from its previous refusal to endorse "lenient" production cutbacks, Iraq gave its assent to the November 4 decision. A detailed breakdown of production in the aftermath of this conference is provided in Table 16. Toward the middle of November, in order to bring that month's allowable volumes in line with the decision taken at the November 5 meeting, some states readjusted their production cutbacks. Table 16 also shows the revised estimate of cutbacks as a result of these readjustments.

The most telling effects of the oil embargo emerged in the form of heightened tensions between the United States and its European allies and Japan involving their respective policies with respect to the Arab/Isrraeli war of 1973. The United States was critical of its allies for not supporting Israel, and the allies expressed their disgruntlement at the failure of the United States to consult with them before embarking on massive military assistance to Israel. The allies quite correctly perceived U.S. military assistance to Israel as a decision of immense magnitude, having a direct bearing on their economic interests in the Middle East. According to a source privy to behind-the-scene negotiations at the October 17 conference of the Arab oil states in Kuwait, until announcement of the American decision to send a large arms shipment to Israel, no firm decision was taken to impose a total embargo on any country. The participants recommended that the United States be subjected to drastic cutbacks from every exporting country, leading to a complete stoppage in Arab supplies. The reason for treating this action against the United States as a recommendation and not a decision was that Saudi Arabia wanted Washington to reconsider its position. When it was known that President Nixon had submitted a request for $2.2 billion worth of military assistance to Israel on October 19, the Saudis retaliated by halting all exports to the United States and by imposing the previously discussed partial reductions of supplies to Europe and Japan.[17] Clearly, the European countries experienced considerable hardship from the oil embargo, as is evident from Table 17. Another aspect of intra-alliance dissension was witnessed in the European Economic Community. As noted in the last chapter, of all the members of EEC, Holland bore the brunt of the oil embargo by losing all its supplies from the Arab states, the equivalent of 71 percent of its total daily imports. A complete breakdown of crude imports to Holland at the time of the

Table 16. Estimate of Arab Oil Cutbacks of November 5 and November 15, 1973 (1,000 b/d)

Producer	Sept. Output	November 5			November 15	
		Cutback	% of Cut	New Output	Cutback	Est. New Output
Saudi Arabia	8,290	2,630	31.7	5,660	2,092	6,198
Kuwait	3,200	950	30.0[a]	2,250	950	2,250
Iraq	2,000	500	25.0	1,500	—	2,000
Abu Dhabi	1,400	350	25.0	1,050	285	1,115
Qatar	600	150	25.0	450	150	450
Neutral Zone	580	145	25.0	435	145	435
Libya	2,300	575	25.0	1,725	575	1,725
Algeria	1,050	263	25.0	787	263	787
Others[b]	1,050	263	25.0	787	263	787
Total	20,470	5,826	28.5	14,644	4,723	15,747

Source: *Middle East Economic Survey,* Nov. 2 and Nov. 16, 1973.

[a]Kuwait actually based its production cutback on the normal level of 3 million b/d rather than the Sept. level of 3.2 million b/d. The reduction from the former was thus 25% and from the latter 30%.

[b]Egypt, Syria, Bahrain, Dubai, and Oman. Dubai and Oman were not parties to the Kuwait decision but announced embargoes on the U.S. and the Netherlands, respectively. Overall cutback for these countries, taking into account the Kuwait decision, war damage, and the reduction in Dubai from an October well fire, was estimated at approximately 25%.

Table 17. National Emergency Energy Measures, December 1973

Country	Speed Limits Cut	Sunday Driving Banned	Gas Rationed	Heating Oil Allocated	Display Lighting Cut	Working Hours Shortened
Netherlands	X	X	X	X	X	a
Italy	X	X		X	X	X
Britain	X		b	X	X	X
Belgium	X	X		X	X	a
Denmark	X	X		X	X	
Germany	X	X		X		a
Ireland	X		b	X	X	
Luxembourg	X	X		X		
France					X	a

Source: U.S. Congress, House, Subcommittees on Europe and on the Near East of the Committee on Foreign Affairs, Hearings, *United States-Europe Relations and the 1973 Middle East War*, 93rd Cong., 1 and 2 sess. (Washington, D.C.: GPO, 1974), p. 75.

[a]Only for auto industry.

[b]Coupons issued.

embargo is provided in Table 18. In this predicament, Holland turned for help to its Common Market partners, proposing measures for sharing of the Community's resources that would have helped its domestic industry. Fearful of losing their own preferred position, though, the Community's response was "entirely negative."[18]

The total Arab embargo of Holland included an additional misery factor for EEC members. The embargo on Holland was also applied to crude oil traded at Rotterdam, one of the most important trade centers of Europe, irrespective of its ultimate destination. Nearly 60 percent of the production of the Dutch refineries was exported all over Europe by ship or through pipeline. Few other ports in Europe could handle the sizable tankers, not to mention store or refine the crude in the volume that Rotterdam handled. Thus, the embargo on Holland affected the whole delivery system of oil in Europe and exaggerated shortages even in those countries that encountered smaller cutbacks.

As anticipated by the protagonists of the oil embargo, the EEC issued a statement endorsing the United Nations ceasefire resolution, which urged the warring parties in the Middle East to pull back to the October 22, 1973, line and to implement resolution 242 of the UN Security Council.[19] This European gesture did not go unnoticed by OAPEC members, who at their November 18 conference, which Iraq did not attend, decided to exempt all EEC members, save Holland, from 5 percent reductions scheduled for December 1 in appreciation of their political stand. The December cuts, however, were applicable to all

Table 18. Crude Oil Imports to the Netherlands, January-April 1973

Producer	Oil Imports (1,000 b/d)
Main Arab Suppliers	
Saudi Arabia	720.8
Kuwait	414.3
Qatar	121.5
Libya	86.5
Abu Dhabi	79.0
Subtotal	1,422.1
Other Arab Suppliers	
Syria	22.7
Egypt	16.2
Algeria	6.4
Iraq	1.9
Dubai	1.0
Oman	0.1
Subtotal	48.3
Total Arab Suppliers	1,470.4
Non-Arab Suppliers	
Iran	362.7
Nigeria	206.8
Gabon	14.9
Venezuela	4.5
Subtotal	588.9
Total	2,059.3

Source: *Middle East Economic Survey,* Oct. 26, 1973.

countries, including EEC. The status of the embargo against the United States and Holland remained unchanged.

In late November, the Arab stance on the oil embargo was hardening. There was a general gloom in Western Europe and Japan over their inability either to put pressure on the United States as desired by the Arabs or to persuade the Arab oil states to soften their position on the embargo. The United States expressed its outrage over what it perceived as pervasive allied "appeasement" in the wake of the Arab "blackmail" of the West. Even though they originally expressed satisfaction with the November 4 statement of the Common Market countries, OAPEC members now felt that the Europeans must further demonstrate their support for the Arab cause by taking substantive actions, such as providing military assistance and applying political pressure on the United States to bring about Israeli withdrawal. Similar pressure was put on Japan, which issued a statement on November 22 along the same line as the one issued by the EEC countries on November 4. But, as one of the world's leading economic powers, Japan could have best served the Arab objectives by striking a crippling blow to the Israeli economy. To this end, the Arab oil states pressed Japan to sever diplomatic and military ties with the Jewish state before they would include Japan in their list of most favored nations. Frustrated over the growing disarray in the Western alliance, the United States threatened to use "counter measures" against the oil embargo. Saudi Arabia's response to this was that if threatened by military action, it would blow up oil installations, thereby denying oil to Europe and Japan for several years.

The Phased Termination The shaky ceasefire between the Egyptian and Israeli forces, which was put into effect on October 22, turned out to be a prelude to the eventual lifting of the oil embargo. This ceasefire was followed by intensification of diplomatic activities not only to bring about disengagement of hostile forces but also to find political solutions which would enable the moderate Arab oil states to end the embargo without losing face. The acceptance of ceasefire by Syria on October 29 was another positive step toward peace. The war of 1973 produced a flurry of diplomatic activities by the United States with the major warring parties as well as with the chief protagonists of the oil embargo. But the tangible results of these activities in the form of favorable changes in observance of the embargo were not immediately noticeable. In fact, all the decisions taken at the Sixth Arab Summit Conference in Algiers, November 26-28, 1973, were harsh toward the West. The 5 percent progressive cuts in Arab oil production were left intact. The participants decided, however, that the loss of revenue for any country as a result of these reductions should not exceed more than one-quarter of its 1972 earnings. Japan and the Philippines were exempted from December cutbacks only. The OAPEC states established a committee to reclassify consuming countries as friendly, neutral, or hostile on the basis of their "stands" and "actions" concerning the Arab cause. An all-Arab embargo was also continued on South Africa, Rhodesia, and Portugal. The oil ministers of Saudi Arabia and Algeria were designated to go to the Western capitals to explain the Arab point of view.[20] As a protest against Egypt's acceptance of the ceasefire, Iraq and Libya boycotted this conference.

The next OAPEC meeting, held December 24-25, 1973, was a marked contrast from the preceding one in that it saw the beginning of a graduated winding down of the embargo. The oil states announced a 10 percent increase in their production, thus leaving their cutback at 15 percent of the September 1973 level. This decision was aimed at improving the supply situation for most European countries, but no changes were made concerning the United States and the Netherlands. OAPEC also singled out Belgium and Japan for preferential treatment. Even though Japan did not break diplomatic ties with Israel as

Table 19. Estimates of Arab Oil Cutbacks of January 1, 1974 (1,000 b/d)

Producer	Sept. 1973 Output	Cutback	Jan. 1 Output
Saudi Arabia	8,290	1,243	7,047
Kuwait[a]	3,200	650	2,550
Iraq	2,000	—	2,100
Abu Dhabi	1,400	210	1,190
Qatar	600	90	510
Neutral Zone	580	87	493
Libya	2,300	345	1,955
Algeria	1,050	157	893
Others	1,050	157	893
Total	20,470	2,939	17,631

Source: *Middle East Economic Survey,* Dec. 28, 1973.

[a]Kuwait based its production cutback on the normal level of 3 million b/d rather than the Sept. level of 3.2 million b/d.

anticipated by the Arabs, it made definite overtures concerning industrialization of the Arab countries.[21] The estimated level of production after this conference is presented in Table 19.

OPEC's decision to introduce quantum leaps in the prices of crude oil at the conclusion of its December 22-23, 1973, meeting in Teheran may very well have been the reason for the decision taken by OAPEC members to reduce production cutbacks on December 25. Saudi Arabia played a vital role in moderating Iranian demands at the OPEC meeting, then publicly expressed dissatisfaction with even the "compromise," which raised the price of Arabian Light marker crude by almost 230 percent from its October 16 level. This increase was to be effective January 1 (see Chapter 4). The Saudi role at the OPEC and OAPEC meetings is instructive of the dilemmas of power faced by that country. While the Saudi regime was resolutely sustaining the political posture of the oil embargo, it was clearly avoiding pushing the Western industrial nations beyond the threshold at which military sanctions against the Arab oil states would become economically enticing.[22] Moreover, the overriding economic interests of Saudi Arabia, as owner of the world's largest oil reserves, demanded that it not preside over such deterioration of the Western economies that their prolonged recovery would endanger the growth of the Saudi economy itself.

Toward the end of 1973, conditions in the Western industrial economies were rapidly deteriorating, and the concomitant impact of the oil embargo in tandem with quantum leaps in crude prices promised to prolong the economic slump through 1974. Economists, businessmen, and government officials in the United States and overseas agreed that even if the Arab states eased their oil cutback, as was evident from OAPEC's December 25 decision, the most optimistic economic outlook for 1974 was worldwide "stagflation," i.e., stagnant or sluggish growth plagued by severe inflation. Concerning the prospects for growth of the American economy, pre-embargo forecasts all predicted a reduced but respectable growth for 1974, but post-embargo forecasts drastically lowered these prospects to near-zero growth or outright recession. For Europe, economists and businessmen almost uniformly saw a business slowdown, even before the embargo. In the aftermath of the embargo, they predicted that the energy crunch could easily transform the slowdowns into recession. Even without considering the impact of the crude shortfall, German economic analysts expected their country's real GNP to rise only 2.5 percent in 1974, down from about 5.5 percent in 1973. France, Britain, Italy, and other continental nations were issuing similar forecasts.[23] Japan was more dependent on Arab oil than the Western European countries and was less prepared than almost any other major oil consumer to deal with a shortage. Within a month after imposition of the oil embargo, off-the-cuff estimates suggested a broad range of production cuts: 6-8 percent in Japan's steel industry, which was only marginally dependent on oil; 10 percent in cement; 12-14 percent in aluminum; and as much as 30 percent in paper and pulp. These cuts, in turn, were expected to affect secondary industries such as motors and electronics in both raw material and power supplies, thus gravely reducing the overall volume of industrial output. The magnitude of "oil shocku" on the Japanese economy was apparent in government estimates that overall economic growth for the remainder of the financial year (through March 30, 1974) was expected to be zero or even negative in comparison to pre-embargo estimates of 9-10 percent growth.[24]

On January 18, 1974, an agreement on the separation of Egyptian and Israeli forces was signed as a result of the "shuttle diplomacy" of Secretary of State Henry Kissinger. Impressed by what he perceived as a new element of substantial evenhandedness in the American foreign policy, President Anwar Sadat of Egypt not only promised Kissinger a relaxation in the oil embargo but also began a campaign to lift the embargo against the United States altogether. His endeavors, however, were met with public opposition by Iraq and Libya, while Kuwait and Saudi Arabia received them without much enthusiasm. The Saudis did not want to take any decisive action on the embargo until a Syrian-Israeli disengagement was negotiated, having given assurance to that effect to President Hafez al-Assad of Syria.[25] The Arab summit held February 14-15, 1974, decided to continue the embargo on U.S. oil supplies. King Faisal, though reportedly sympathetic to Sadat's advocacy of suspension of the embargo on the United States, was not completely satisfied that the latter had applied enough pressure on Israel to withdraw from Arab territory occupied in 1967.[26]

No announcement to terminate the oil embargo was issued at the conclusion of the OAPEC meeting in Tripoli, Libya, March 13-14, 1974, but a

Table 20. World Crude Production, September 1973-March 1974 (1,000 b/d)

Source	Sept.	Oct.	Nov.	Dec.	Jan.	Feb.	Mar.
Arab	20.8	19.8	15.8	16.1	17.6	17.9	18.5
Non-Arab	38.4	38.9	39.0	39.3	39.6	39.5	39.5
Total	59.2	58.7	54.8	55.4	57.2	57.4	58.0

Source: *Oil and Gas Journal*, Aug. 31, 1974.

decision to do so was a foregone conclusion by this time. The intensive diplomatic activities of Henry Kissinger created hope that meaningful progress on the Arab/Israeli conflict was being made. And the Saudis expressed their willingness to end the embargo. On March 14, 1974, almost five months after the October 17 decision to impose the oil embargo, Saudi Arabia, Kuwait, Abu Dhabi, Qatar, Bahrain, and Egypt lifted the embargo against the United States "indefinitely"; Algeria agreed to lift it for two months only; and Libya and Syria refused to end the embargo at all. The embargo against Holland continued. OAPEC members were to meet again on June 1 to review the situation. For all practical purposes, however, the Arab oil embargo was about to become history.

As a result of the embargo, world crude oil production declined by 5 percent between November 1973 and February 1974. This decline, which reflected a cutback of 19 percent in Arab production, was alleviated in part by a 2 percent increase in non-Arab oil production. Table 20 shows world crude oil production from September 1973 through March 1974.

The main thrust of this chapter has been that even though the decision to impose the oil embargo was aimed at the attainment of political objectives, namely, creating a negotiating momentum toward resolution of the Arab/Israeli conflict, the oil embargo itself was very much in the economic interests of its implementors. The fact that the embargo was not in any way causing a loss in the oil revenues enhanced its utility as a weapon in the hands of the Arab oil states for extracting political concessions from the industrial consuming nations. It clearly failed to achieve its stated objective of bringing about Israeli withdrawal from occupied Arab territory. But it was a remarkable success in bringing an end to the lengthy impasse involving that conflict.

The oil embargo pushed the supply- and price-related uncertainties of the sellers' market to new heights. Its most significant outcome was that it firmly established the Arab oil states as advantaged actors. Since 1971, these states had been in the vanguard of OPEC's activism without, perhaps, being manifestly identified as such. Libya, for example, performed the groundbreaking task of concluding the Tripoli I agreement, a performance repeated in Tripoli II. The

Table 21. Oil Consumption in Major Consuming Areas, 1973-1974 (million of tons)

Country/Region	Jan.- April 1973	Jan.- April 1974	% of Change
United States	220.9	205.8	−6.8
Japan	81.5	82.4	+1.0
Western Europe[a]	152.2	135.4	−11.0
Canada	26.4	28.1	+6.0
Total	481.0	451.7	−6.3

Source: *U.S. Oil Companies and the Arab Oil Embargo: The International Allocation of Constricted Supplies*, report prepared for the Subcommittee on Multinational Corporations of the Senate Committee on Foreign Relations by the Federal Energy Administration, Office of International Energy Affairs (Washington, D.C.: GPO, 1975), p.8.
[a]United Kingdom, France, Germany, and Italy.

Teheran agreement was probably one of the few agreements in which Iran shared the limelight with Arab oil states such as Saudi Arabia, Kuwait, and Iraq. The credit for concluding the participation agreements goes largely to Saudi Arabia, even though Libyan, Iraqi, and Algerian tactics made important contributions in signaling the oil companies that the alternative to rejection of the Saudi participation formula would be outright nationalization. In the aftermath of the embargo, however, the Arab oil states emerged as a force that could not be easily dismissed or taken for granted by either the West or the East.

Immediately after imposition of the embargo, critics contended that the Arabs could not stop the major oil companies from supplying oil to the United States. The efficient planning of the embargo surprised both friends and foes of the OAPEC nations. Although there were some reported "leaks" to the embargoed nations, considering that oil is a nontraceable commodity and especially that swapping arrangements between the various integrated oil companies is very much standard operating procedure and a long-standing matter of convenience, the amount of reported leakage was remarkably low.[27] Three factors disciplined the behavior of the companies. The first was Saudi Arabia's efficiency in tracking the movement of its commodity through international shipping lanes. The second was the penalty for violating the rules of the game established by the Arab oil states. The third deterrent was the repercussions from the industrial consuming nations in the oil companies' appearing to be sycophantical to the OAPEC countries. Consequently, according to one report, in the distribution of constricted supplies, the five major American oil companies—Exxon, Socal, Texaco, Gulf, and Mobil—treated the United States and Western Europe "quite evenly." The United States was marginally better off, receiving 12 percent less in crude and other products, while Europe received 13.6 percent less. Japan, on the other hand, enjoyed a 1 percent rise in imports during the

Table 22. Deliveries of Crude Oil and Products by Five Major U.S. Oil Companies, 1972-1974 (1,000 b/d)

Country/Region	Base Period (Dec. 1972- March 1973)	Embargo (Dec. 1973- March 1974)	% of Change
United States	2,607.3	2,294.8	−12.0
Canada	506.1	474.6	−6.3
Western Europe[a]	4,859.3	4,198.3	−13.6
Japan	1,683.3	1,701.0	+1.0
Other	2,795.3	2,918.1	+4.4
Total	12,451.3	11,586.8	−6.9

Source: *U.S. Oil Companies and the Arab Oil Embargo: The International Allocation of Constricted Supplies*, report prepared for the Subcommittee on Multinational Corporations of the Senate Committee on Foreign Relations by the Federal Energy Administration, Office of International Energy Affairs (Washington, D.C.: GPO, 1975), p. 11.

[a]United Kingdom, France, Germany, and Italy.

embargo. This discrepancy was based on the fact that during 1973, Japanese energy demand was growing at approximately 17 percent while America and Western Europe were experiencing a far more modest growth rate of approximately 5 percent.[28] Tables 21 and 22 present a comparison of oil consumption and deliveries to major industrial consuming areas for 1973 and 1974. Several factors should be noted with regard to these tables: (1) the base period differs from the base used for production statistics; (2) the higher petroleum prices in Western Europe tended to curtail demand more than in the United States, where price levels remained significantly lower; (3) Japanese demand had been growing much more rapidly than American or European demand; and (4) Canada, as both a petroleum-importing and a petroleum-exporting nation, occupied a unique position.[29]

Some Arab oil loaded before the embargo was landed in the United States during the embargo, and there was some leakage from the smaller Arab producers. But by and large, the embargo did choke the flow of Arab oil to the United States. To compensate, the major international corporations imported larger quantities of nonembargoed oil from sources such as Iran (an increase of 41.8 percent over the pre-embargo base period), Nigeria (up 66.9 percent), Indonesia (up 28.8 percent), Venezuela (up 3.3 percent), and other Caribbean sources (up 131.4 percent).[30] The Arab oil embargo also caused considerable damage to the U.S. economy. According to one source, the output of the economy in the first quarter of 1974 fell by $10 to $20 billion as a result of the embargo. The embargo is estimated to have caused about 0.5 percent reduction of the civilian labor force, or job losses for approximately 500,000 people. The automobile industry was the hardest hit area of the industrial sector. The U.S.

Department of Labor estimated that almost 80 percent of the industrial layoffs were attributed to energy problems and could be traced to the decline in demand for automobiles or recreational vehicles. Throughout the embargo period, the Midwest accounted for approximately two-thirds of all energy-related unemployment, Michigan accounting for upwards of 70 percent. Energy prices during the embargo period were responsible for at least 30 percent of the increase in the consumer price index.[31]

Saudi Arabia enormously enhanced its prestige through its skillful separation of the embargo from the issue of prices for crude oil. As a member of OAPEC, Saudi Arabia insisted on sustaining the embargo against the United States until convinced that the U.S. was genuinely applying pressure on Israel for military disengagement and resolution of the Arab/Israeli conflict. At the same time, as a member of OPEC, Saudi Arabia refused to go along with unrealistic escalations in the prices of crude oil lest its own economy become the victim of such precipitous increases. The insistence on the Israeli withdrawal from Arab territory as a precondition for lifting the oil embargo was subtly dropped when the Saudi regime became convinced of the magnitude of American involvement in negotiating settlement of the conflict. Such nuances and flexibility in Saudi foreign policy guaranteed that country a major role not only in setting the future parameters of the negotiated peace process but also in determining the future dimensions of the superpower strategic balance in the Middle East.

Iraq was the only consistent nonparticipant in the Arab oil embargo. The most obvious reason appears to have been economic, even though this rationale remained shrouded in massive political rhetoric. As a result of a continuing dispute between the government and the foreign oil companies, Iraqi oil production remained rather erratic and largely depressed. Table 13 shows that Iraq was the only Arab country that lost production and consequently revenues not as a result of deliberate policy, which was the cause of production and revenue losses in Libya. Full-fledged Iraqi participation in the oil embargo would probably have placed further strain on the government's ambitious development programs. The Iraqi government fulfilled its political commitment to the Arab cause by implementing a total embargo on American and Dutch supplies. These measures, which resulted in only minor and symbolic cutbacks, saved Iraq from the economic hardships that full participation in the oil embargo would have caused. Tables 14, 16, and 19 show actual production cutbacks for Iraq.

The long-term effects of the oil embargo were more psychological than real. Its significance is not that it could be repeated in the foreseeable future but that, like any political or economic disaster—Pearl Harbor, a war, the Great Depression—it serves as an uneasy reminder to the Western economies of the consequences of lack of foresight, of unplanned plundering of natural resources, and of reckless dependence on sources of supply that could be eliminated by design or by accident. For the future, however, it is comforting to note that the West is much better prepared for an embargo-type exigency. In addition, Saudi Arabia, which was the pacesetter for the 1973 embargo and whose participation would be a vital precondition to a new embargo, is too dependent on the West for economic development and technology transfer and too preoccupied with issues of regional security to be tempted by such an action. Finally, and most importantly, a future embargo may no longer be economically beneficial to its sponsors simply because it would be bound to drive the price of OPEC oil to a range of $50- $70/b, at which point oil might lose its competitive advantage over alternative fuels.

The period of the oil embargo may have been the first time when the economies of the oil states were at least temporarily secure from the effects of international inflation. This security was due largely to their enormous current account surpluses, as noted in the last chapter. It was between 1975 and 1978 that the oil states finally faced the grim consequences of their quantum price escalations of October-December 1973. During those years the industrial economies strove to bounce back from the shocks of 1973, and the oil states struggled to cope with increasing inflation and erosion in their buying power and to minimize downward pressure on their prices. A detailed examination of these phenomena will occupy us next.

6. Toward an Uncertain Future

From 1976 through 1978, the oil states continued to suffer from the "catch-22" effects of their quantum price increases of the last quarter of 1973. These effects included the dwindling of their oil surpluses, the erosion of their buying power, and, most important, the continued downward pressure on the prices of their crude oil. In the latter part of 1978, the cataclysmic political change in Iran intensified the uncertainties of the world oil supply, and the oil states once again exploited the situation by introducing excessive price increases through 1980. From 1981 through 1983, while the industrial countries were recuperating from the economic slump, OPEC was desperately scrambling to safeguard not only its price scale but also its production volumes. The oil states even tried their hand, unsuccessfully, at incorporating production programing in 1982. In March 1983, OPEC adopted another production programing arrangement in tandem with—for the first time in its history—lowering its prices. The March 1983 decision to reduce production was repeated in October 1984. When this action failed to forestall downward slides in oil prices, OPEC reduced the prices of its crudes once again in January 1985. Despite these measures, the future of the organization appeared uncertain at the conclusion of the first quarter of 1985.

Four issues were of primary importance during this period. The first was the ability of Saudi Arabia to act as a "swing" producer (i.e., to raise or lower production rates significantly without loss to its economy) and the influence of this ability on the pricing behavior of other OPEC producers.[1]

The second issue was the pricing behavior of OPEC during the shortages that resulted from the political revolution in Iran. When the dissarray in the oil market of 1979 and 1980 dissipated, spot prices, not OPEC, set the pace for future prices. OPEC's quest for a long-range strategy with the manifest purpose of stabilizing the industry under a calm market was also set aside.

The third issue of concern was the behavior of OPEC in the oil market of the 1980s. A variety of factors, such as recession, the institutionalization of con-

Table 23. OECD Output and Price Forecasts, 1974-1976 (% of change)

Country/Region	Real GNP			Consumer Prices		
	1974-1975	1975-1976	1975[a]-1976[a]	1974-1975	1975-1976	1975[a]-1976[a]
United States	−3.00	+5.75	+4.75	+8.00	+7.00	+7.00
Japan	+1.25	+4.25	+5.00	+12.25	+9.75	+10.00
OECD Europe	−2.50	+2.00	+2.50	+12.00	+9.50	+9.25
Total OECD	−2.00	+4.00	+4.00	+10.50	+8.50	+8.50

Source: *OECD Economic Outlook*, Dec. 1975.

[a]Second quarter.

servation and fuel switching in the OECD economies, and the increasing influence of non-OPEC producers, resulted in transforming OPEC into a residual producer.

The final issue was intra-OPEC tensions. Between 1976 and 1978, these tensions were primarily economic, but after 1980 the organization saw the emergence of political tensions as well. These involved the fire-brand revolutionary regime of Ayatollah Rouhollah Khomeini in Iran and the Bathist Iraqi regime of Saddam Hussein, which have been fighting a war since September 1980. The conservative Arab monarchies of the Persian Gulf have supported Iraq, while Libya has sided with Iran. Another growing political tension has developed between Iran and Saudi Arabia for leadership of OPEC. These particular tensions have been around since the days of the Shah, and until 1979 they were manageable. Under Khomeini, Iran has continued to try to retain primacy within OPEC. This desire, in combination with the continuing Iran/Iraq war and the growing political tension between Iran and the conservative oil monarchies, is likely to incapacitate OPEC from within at a time when this organization is trying to cope with growing external challenges from the continuing economic downturn in the industrial world.

The Two-Tier Pricing Agreement

In late 1975 and early 1976, the Western industrial economies were trying to pull themselves out of the great recession. The American economy was moving up, very much along the forecasts made in early 1975, but the recession bottomed out before the middle of that year. In Japan, falling output was followed by mild expansion, and the government adopted measures to reinforce the recovery. In Europe, despite a number of measures taken by various governments to strengthen demand, the rate of recovery was slower than had been anticipated. The result of all this was that the recession was in the process of dissipating. The most striking aspects of the recovery were its slowness (by the standards of earlier recovery periods) and its failure to gather strength in the course of 1976.[2] A summary of output and price forecasts for this period, provided in Table 23, substantiates the upswing in economic activity in the OECD countries. The world demand for crude oil, however, was still suffering from the recession and from high prices, which made fuel conservation a way of life and switching to alternate fuels an enticing option for the Western economies. This slump in demand had left the oil states with excess capacity. Past premiums (i.e., differentials) imposed by the oil states during a tighter market could not be maintained. At their June 1975 conference in Libreville, Gabon, the OPEC members attempted to devise mechanisms aimed at bringing about automatic adjustments in quality and freight premiums. But since in the current soft market any mechanisms governing differentials would have necessitated reductions in quality and freight premiums, the oil states decided to postpone the adoption of such mechanisms. In 1976, and under prolonged soft-market conditions, the issue of differentials continued to cause minor irritations among OPEC members. For example, in December 1975, Algeria accused Iraq of selling its Mediterranean crude for 30-40¢/b less than it should have in relation to Arabian Light marker crude, and 10-15¢/b less than it should have under a strict application of the 10 percent increase decided on in October. Algerian irritation at the Iraqi pricing policy was also apparent in the former's abrupt rejection of the Iraqi call for a special meeting of OPEC ministers to discuss differentials. Iraq itself was critical of an agreement concerning differentials reached in October 1975 between Saudi Arabia and Kuwait, which resulted in a downward adjustment in the prices of Kuwaiti and Saudi heavy crudes. In February 1976, Iran, in response to the prolonged slump in its production of heavy crude, which was considered overpriced, announced a reduction of 9.5¢/b, from $11.495 to $11.40/b. Even this cut was not expected to bring about any significant increase in the demand for heavy crude. In fact, in view of the sustained lower demand for heavy crude, Saudi Arabia announced the closure of two of its oil fields whose combined output amounted to about half of that country's heavy crude production. It is clear that under continued soft market conditions, OPEC was increasingly hard pressed to take preventive measures concerning differentials. In April, Algeria offered a complicated formula for developing single composite differentials for each crude, a formula expected to be the basis for determining differentials between the prices of various OPEC crudes in the future.

Toward the end of the first quarter of 1976, U.S. oil imports made a record upward jump, the result of an escalating demand for petroleum products and a continued decline in domestic production of crude.[3] In January and February, average U.S. imports were reported to be 7.8 million b/d, as opposed to 6 million b/d in the same months of the preceding year, and in March this average

Table 24. World Oil Production, 1975 and First Half of 1976

Region	Production, 1975		Production, First Half of 1976	
	(1,000 b/d)	% of change	(1,000 b/d)	% of change
North America				
Canada	1,444	−14.5	1,280	−7.2
U.S.	8,351	−4.7	8,141	−3.7
Latin America	4,277	−11.3	4,178	−5.7
Europe	536	+39.9	714	+51.6
Africa	4,951	−7.4	5,549	+23.9
Middle East	19,569	−10.00	20,085	+4.7
Asia/Pacific	2,200	−4.7	2,425	+13.7
Communist areas	11,711	+8.8	12,192	+6.4
Total	53,051	−5.4	54,563	+5.0

Source: *Oil and Gas Journal*, Feb. 16, 1975, Aug. 30, 1976.

further shot up to 8.2 million b/d.[4] The increased demand in the United States was a significant aspect of the overall increase in world demand for crude oil that was anticipated at the beginning of 1976. Observers of the oil scene made various estimates of OPEC recovery potential for that year, ranging from a possible growth of less than 1 million b/d to full restoration to pre-1975 levels of nearly 31 million b/d.[5] Toward the end of the first six months of 1976, worldwide crude oil production indicated definite and sustained signs of recovery. Table 24 shows regional oil production for 1975 and the first half of 1976. The rapid expansion in the United States since mid-1975 provided a considerable boost to recovery in other countries, especially Japan, OECD industrial production was close to the peak reached in 1973.[6] Against this background, price moderate Saudi Arabia and price hawks Algeria and Venezuela initiated preconference lobbying for their respective positions.

After stormy debates between moderates and hawks at an OPEC meeting held May 27–28, 1976, in Bali, Indonesia, the hawks reluctantly consented to the Saudi insistence on continuing the price freeze on Arabian Light marker crude at $11.51/b until the end of 1976. The decisive factor was the Saudi threat to increase production unilaterally and to maintain a lower tier of prices than other OPEC states. Saudi Arabian opposition to a price increase was based on fear that precipitous escalations at this juncture might serve as a death blow to the emerging world economic recovery. On the issue of differentials, no long-term decision was made. The selling prices of medium and heavy crudes in the Gulf area were reduced by 5-10¢/b as a temporary measure. The prices of lighter crudes were left for the individual governments to decide, with the understanding that these prices would not be less than prevailing levels. The fact was that, with the overall upsurge in the demand for lighter crudes, the issue of differentials was no longer contentious. Even the price scale for Iraqi Mediterranean crude, a source of friction between Algeria and Iraq, dissipated when the latter ceased its exports to the eastern Mediterranean because of a transit dispute with Syria.[7]

In the third quarter of 1976, worldwide crude production was on the upswing. It was reported that world output during the first half of 1976 averaged nearly 54.6 million b/d, a gain of 5 percent. The non-Communist production averaged nearly 42.4 million b/d, for a half-year gain of 4.5 percent. OPEC oil flow rose by 8.2 percent, to more than 29.9 million b/d, with June production nearing the 30 million b/d mark. But this increased output was still below the pre-embargo OPEC peak of 32 million b/d in September 1973. The estimated production of OPEC during the third quarter of 1976 was about 38 million b/d.[8] According to extrapolations made in an authoritative study, however, the increasing growth of the world economy was expected to create a market for 34.2 million b/d of OPEC crude by mid-1977, a substantial increase from the June 1976 rate of 29.9 million b/d.[9] In July, Saudi Arabia reported its third consecutive increase in oil production, temporarily exceeding by .2 million b/d its previous self-imposed ceiling of 8.5 million b/d.[10] In addition to the improved supply situation, prices of light crude were marked by steady increases. Libya raised the price of its light crude by as much as 19–20¢/b. Some quality crudes in Nigeria and Algeria were being sold at $13/b or more. Under these substantially improved market conditions, the OPEC price hawks, who were clearly unhappy with the Bali decision of a continued freeze, were expected to hold out for a 10-15 percent boost in the prices of crude oil at their next scheduled meeting in December. The impatience of the price hawks became apparent in frequent calls for an Extraordinary Meeting to discuss price increases before the December meeting. Saudi Arabia, using the same argument for a price freeze as at the Bali conference, refused to go along with such suggestions.

As OPEC's December meeting drew closer, reports of preconference jockeying for position became more widespread. The price hawks sent up trial balloons of their demand for a price boost of 25 percent, while the price moderates reportedly reached an informal understanding to opt for a 10 percent increase. For the Saudis, however, two additional factors were to play significant roles in the price increase they would allow at the December conference. The first factor was the outcome of the North-South conference. At a press conference in Geneva in August, Yamani stated that the Saudi decision concerning the extent of price increase at the December OPEC meeting would be influenced by progress in the impending North-South dialogue. The second factor was the election of Jimmy Carter as the president of the United States. As the president-elect was preparing himself for the highly intricate business of governing America and handling the intractable problems of the international economy, Saudi Arabia, as a gesture of goodwill toward him, opted to press for moderate price increases. The Saudi hope was that, as a quid pro quo, Carter would create new momentum toward settlement of the Arab/Israeli conflict to the satisfaction of the Arabs and would show good faith toward speedy progress in the North-South dialogue.

The Paris Conference on International Economic Cooperation (CIEC, or the North-South conference) was scheduled to culminate with a ministerial meeting on December 15-17. The OPEC conference was originally planned to be held on December 15 in Doha, Qatar. But, because a majority of the oil states, including Saudi Arabia and the Third World allies, wanted to link progress at the North-South conference with the next OPEC price increase, there were efforts to postpone the OPEC conference to December 20. This shuffling of dates turned into a power game in which both the industrial countries and the Third World bloc tried to guess the reciprocal linkage between their concessions and those of the other side. The industrialized states wanted to be sure that their willingness to offer concessions of commodity prices and debt relief and to conclude the dialogue on schedule would be reciprocated by the oil states in the form of moderate price increases at their next meeting. The OPEC/Third World countries, on the other hand, wondered if their willingness to postpone the Paris ministerial conference in tandem with moderation on the price issue in Doha would lead to the desired economic concessions by the West. During the first week of December, it became apparent that the CIEC ministerial conference would not take place on December 15-17. Nevertheless, the OPEC moderates decided to maintain their advocacy of a maximum 10 percent price increase at their December conference, which was to be held as originally scheduled. This decision was based on acceptance of the International Monetary Fund's index of export prices of industrial countries, which had remained steady since the third quarter of 1975. The price hawks, on the contrary, using data supplied by OPEC's Economic Commission, were demanding increases as high as 30 percent. According to the Economic Commission, increases in the prices of OPEC imports since the last oil price increase in October 1975 had been between 20 and 30 percent.[11]

Toward the end of November, the industrialized countries mounted a diplomatic campaign aimed at persuading the price moderates to forgo price increases in December. The United States initiated behind-the-scenes consultations with members of OECD aimed at concerted action to avert the increases. Estimates of possible damage to the industrial economies from OPEC price increases were also released to dissuade the moderate oil states from conceding to the demands of the price hawks. It was estimated, for example, that each 1 percent advance in the world oil price would add $1 billion to the oil bill of the big seven countries—the United States, Japan, West Germany, France, the United Kingdom, Italy, and Canada—and would translate into a 0.5 percent rise in the inflation rate for larger imports. The impact on the American economy of a potential 15 percent increase was estimated to be an additional payment of $5.5 billion, an escalation of more than a percentage point in the prevailing 6 percent inflation rate, and a .06 percent reduction in the prevailing economic growth rate of 4 percent. A study released by OECD, which was interpreted as an attempt to persuade OPEC to moderate its price position, stated that the expected revised growth for 1977 would be 4.3 percent for the first half and 5 percent for the second, as compared to its earlier estimates of 5.25 percent for the first half and 5 percent for the second. Chancellor Helmut Schmidt of West Germany, British Prime Minister James Callaghan, and President-elect Jimmy Carter also made several public statements urging OPEC to refrain from untimely and unwarranted price increases.[12]

The collective pressure applied by the Western industrial nations produced its intended results even before the OPEC conference. Saudi Arabia stated that it favored a continuation of the price freeze for another six months. At the December OPEC conference in Doha, the outcome of the industrial consumers' campaign was startling. The position of the price hawks on the amount of increase varied from a low of 10 percent (Kuwait, Venezuela, and Indonesia) to 15 percent (Iran, Nigeria, Libya, Algeria, Qatar, Gabon, and Ecuador) to a high of 26 percent (Iraq). Saudi Arabia, supported by the UAE, advocated a continuation of the price freeze for another six months. In a radical departure from their past tradition of arriving at a compromise formula at the last moment, neither the price hawks nor the Saudis were willing to agree on a uniform price increase. In view of the acute variance between positions, both sides sought solace in "agreeing-to-disagree," and this "consensus" gave birth to a two-tier price system. According to this arrangement, eleven oil states agreed to bring about an approximately 10 percent increase in their prices from January 1, 1977, and an additional 5 percent at midyear, while Saudi Arabia and the UAE announced a 5 percent increase for the entirety of 1977.[13] As a tactical maneuver aimed at applying pressure on the price hawks, Saudi Arabia and the UAE announced the removal of their previous self-imposed production ceilings, thereby raising their combined production from the prevailing rate of 11.1 million b/d (9.1 million for the Saudis and about 2 million for the UAE) to 12.5 million b/d (about 10 million for the Saudis and about 2.5 million for the UAE), or about 42 percent of the total OPEC production, which remained at 32-33 million b/d. The Saudis hoped that the surplus thus created would produce downward pressure on the price scales adopted by the eleven other oil states and that the latter would be forced to realign their prices with those of Saudi Arabia and the UAE.[14] The OPEC majority was equally determined to back its higher price tier even with coordinated production cutbacks, if necessary, rather than succumb to the Saudi pressure.

As the realities of maintaining a two-tier price system were beginning to sink in, both price hawks and moderates were becoming increasingly aware of the practical difficulties of adhering to their respective positions. The Shah of Iran, in the wake of the Saudi estimate that the production of the eleven OPEC price hawks might drop by more than 25 percent because of lower Saudi prices and increased production, stated that any attempt by the Saudis to use escalated production as a pressure tactic would be tantamount to an act of aggression. Iraq, expecting a 30 percent drop in its production, initiated an intensive behind-the-scenes lobbying campaign for an Extraordinary Meeting of OPEC members to reconsider the two-tier price system. Kuwait stated that any Saudi increase in heavy and medium crudes would adversely affect the latter country because Kuwaiti production was made up entirely of medium and heavy crudes. Saudi Arabia, in escalating production, encountered a few problems of its own. In January 1977, Saudi production remained about 1 million b/d short of the projected level of 10 million b/d for the first quarter of that year. For February, the production was 8.76 million b/d, and the output for March averaged at 9.3 million b/d. The underlying reason was bad weather conditions during all three months. Moreover, Abu Dhabi, a member of the UAE, was vehemently opposed to any relaxation of allowable ceilings placed for technical reasons. Then there were questions concerning the estimated spare-producing capacity of the UAE, which turned out to be around 200,000 b/d, much lower than earlier estimates of 300,000 to 500,000 b/d.

When Saudi Arabia and the UAE decided to maintain a lower tier of prices and coupled that action with increased production, they also hoped that their cheaper oil would attract a horde of purchasers. The reality was contrary to their expectation, however. The OPEC price hawks made it clear that they expected the oil companies to honor their commitments by lifting the contracted oil if they wanted to do business with them in the future. The oil companies thus found themselves locked into higher-priced oil in order to protect their long-range sources of supply. An additional factor working against cheaper Saudi crude was that oil producers such as Libya, Algeria, and Nigeria continued to sell their oil at premium prices. Their ligher, low-sulphur crudes remained much in demand, particularly in the United States, where many refineries were not equipped to process heavier high-sulphur Saudi oil. Then, in the middle of May there was a fire at the Abqaiq producing center in Saudi Arabia. This incident, which led to a reported decline of 2 million b/d, dealt a major blow to the sustained Saudi strategy of bringing an end to the two-tier price system on Saudi terms.

Between January and May of 1977, several behind-the-scenes endeavors were in progress to find a pricing formula that would be acceptable to both the hawks and the moderates without causing either any embarrassment. But Saudi Arabia remained cool toward any formula that would raise prices more than 5 percent during 1977. In April, President Carlos Andres Perez of Venezuela undertook a conciliatory mission during his visit to the Persian Gulf oil states.

About the same time, three members of the OPEC majority—Kuwait, Qatar, and Venezuela—which opted for a total 15 percent increase for 1977, decided against applying the 5 percent increase that was due in July. Soon thereafter, Iran and Iraq joined them. This action by the OPEC majority created the possibility of the emergence of a three-tier price structure: one group of countries going for a total increase of 15 percent, another group remaining at 10 percent, and Saudi Arabia and the UAE staying at 5 percent. This situation created what the Saudis saw as the perfect opportunity to take charge by attempting to find cooperative avenues that suited the economic objectives of all parties. After all, the oil market had not been reacting according to Saudi expectations. The upper-tier countries were shielded from any substantial loss of oil by the combination of a strong demand, resulting from severe winter conditions in the industrial consuming areas, and Saudi Arabia's failure to meet its first-quarter output target. The Abqaiq fire prevented the Saudis from escalating their rate of production. In addition, the failure of the North-South dialogue to produce significant results did not encourage Saudi Arabia to maintain a lower price, which was essentially a policy sympathetic to the Western industrial nations.[15]

In an interview with *Business Week*, Yamani confirmed that his country would end the two-tier price system by raising its own prices by a maximum of 5 percent on July 1. At the next conference in Stockholm, July 12-13, 1977, this decision was endorsed by the OPEC membership, thus unifying the price of Arabian Light marker crude at $12.70/b for the remainder of 1977. The two-tier price system caused little damage to the OPEC price hawks. They learned to defy Saudi Arabia despite its role as a swing producer, and the Saudis had to choose another occasion to demonstrate the potential power of their enormous oil reserves as a modifier of the pricing behavior of other OPEC members.

The Calm before the Iranian Turmoil

The second half of 1977 was marked by a surplus of 2 million b/d in the oil market from non-OPEC sources, largely North Sea and Alaskan oil. As shown in Table 26, U.S. production from Alaska jumped from 173,000 b/d in 1976 to 464,000 b/d in 1977. U.S. imports, however, continued to follow an upward pattern, growing from about 5.3 million b/d in 1976 to 6.6 million b/d in 1977.[16] Crude production in Mexico and the United Kingdom made substantial jumps from 1976 to 1977, while Norway recorded no change and Canada little change between these years (Table 25).[17] In addition, petroleum storage tanks in the United States were brimming with 970 million barrels of crude oil and fuel, the reported inventory up 11 percent over the previous year. The economic upsurge of late 1976 and early 1977 was short-lived, however. Total OECD growth, which weakened in the second quarter of 1977, remained sluggish toward the end of the year. Industrial production broadly stagnated after April.

Table 25. Crude Oil Production by Major Non-OPEC Sources, 1973 - March 1984 (1,000 b/d)

										1984[a]		
Producer	1973	1976	1977	1978	1979	1980	1981	1982	1983[a]	Jan.	Feb.	March
Norway	32	279	280	356	403	528	501	520	614	688	680	672
United Kingdom	2	245	768	1,082	1,568	1,622	1,811	2,065	2,299	2,560	2,640	n.a.
Mexico[b]	465	831	981	1,209	1,461	1,936	2,313	2,748	2,684	2,668	2,757	2,711
Canada[b]	1,800	1,295	1,320	1,313	1,496	1,435	1,285	1,372	1,320	1,600	1,600	n.a.

Source: *International Energy Annual 1983*, p. 16.

[a]1983 and 1984 figures are from *International Energy Statistical Review*, June 1984, p. 1.

[b]Mexican production for 1974 and 1975 was 571 and 705, respectively; Canadian production for 1974 and 1975 was 1,684 adn 1,439, respectively.

Table 26. Sources of U.S. Crude Oil Needs, 1973-1984 (1,000 b/d)

Year	Field Production		Total Imports
	Total Domestic	Alaskan	
1973	9,208	198	3,244
1974	8,774	193	3,477
1975	8,375	191	4,105
1976	8,132	173	5,287
1977	8,245	464	6,615
1978	8,707	1,229	6,356
1979	8,552	1,401	6,519
1980	8,597	1,617	5,263
1981	8,572	1,609	4,396
1982	8,649	1,696	3,488
1983	8,688	1,714	3,329
1984	8,753	1,742	3,436

Source: U.S. EIA, *Monthly Energy Review*, Dec. 1984, p. 44.

The U.S. economy was slowing down, and industrial output in Japan was flat. In Europe, output fell sharply in the second quarter and projections of economic activity were not optimistic. World trade growth, which was very modest in the first half of 1977, failed to show marked improvement toward the end of the year.[18]

Despite a sustained soft market resulting from the economic slowdown in the industrial countries and despite frequent American pleas to freeze prices of crude oil, it was expected that the oil states would raise the prices of their crudes by 5-10 percent. The Shah of Iran, however, as a result of his mid-November visit to the United States, decided to support a price freeze for the entire 1978 year. His decision was widely believed to be a quid pro quo for the American readiness to fulfill Iranian defense needs, an element that strengthened the Saudi position. Saudi Arabia was expected to support a price freeze initially at the forthcoming OPEC meeting at Caracas, but was reportedly willing to settle for a 5 percent increase. Now, with the Shah backing a price freeze, the Saudis (who, along with the Iranians, controlled 50 percent of OPEC output) no longer felt the need for a compromise with the rest of the OPEC membership. At the Caracas meeting on December 20-21, 1977, in view of mounting pressure from Saudi Arabia, Iran, the UAE, Kuwait, and Qatar, OPEC decided to freeze the price of Arabian Light marker crude at $12.70/b.[19]

During the first half of 1978, world oil production was down by only 1.1 percent, compared to the 2.3 percent production drop toward the end of the first quarter. The output of the non-Communist area was down by 2.8 percent, compared to 4.2 percent toward the end of the first quarter. OPEC production remained at 28.56 million b/d, a slight increase from 28.2 million b/d at the end of the first quarter, but a significant drop from 31.5 million b/d in the fourth quarter of 1977. The major brunt of this cutback was borne by Saudi Arabia, where production dropped to 7.75 million b/d from a peak of 8.9 million b/d during the fourth quarter of 1977. The primary reason for OPEC production cutbacks was reported to be escalated production in the North Sea, Mexico, and the Alaskan North Slope. The United Kingdom's share of North Sea output topped the 1 million b/d mark in May 1978, its average for the first half of the year remaining at 984,000 b/d. Norway's Ekofisk complex in the North Sea was producing 351,000 b/d for the same period. Mexican output for June was reported to be 1.7 million b/d; its exports were 400,000 b/d and were expected to go as high as 700,000 b/d by the end of the year. The Alaskan North Slope production, which boosted American output to 8.6 million b/d during the first half of the year, remained at 980,000 b/d, effectively shutting off crude imports, mainly Middle East sour crude.[20] It should be pointed out, however, that the competition from non-OPEC sources was not the only reason for OPEC production cutbacks. A large number of OPEC buyers, particularly the United States, the Western European countries, and Japan, were working off huge inventories of crude and refined products piled up from the previous year.[21] This measure would come back to haunt the industrial consumers when the political turmoil in Iran in late 1978 drastically reduced crude production in that country and created a supply crisis in the world crude market.

In general, the response of OPEC members to the oil glut was surprisingly mild. In mid-January 1978, Saudi Arabia reimposed its previous production ceiling of 8.5 million b/d, then, toward the end of February, issued a directive that output of Arabian Light should not exceed 65 percent of total crude oil production in 1978. This decision gave crude oil production in Iran and Iraq a slight boost. Libya, Algeria, and Nigeria also took a variety of measures to ease the mounting pressure from escalated production by non-OPEC sources. Libya and Algeria lowered the prices of their light low-sulphur crudes by 20¢/b, effective January 1, 1978. In April, Nigeria joined their ranks by announcing downward adjustments of 15-23¢/b for most crudes for the second quarter of 1978. Nigeria cut its production from 1.9 million to 1.64 million b/d, and Libya cut its from 2.05 million to 1.89 million b/d. Iran and Kuwait also reduced the prices of their heavy crudes to compete better with comparable Saudi grades.

In the second half of 1978, OPEC members strove to come to grips with two major issues. The first was their age-old concern to formulate a long-range production strategy that would forestall inordinate decreases in oil prices in a soft market. Their second major concern was the impact of the decline in the American dollar on their oil revenues.

The soft state of the oil market made this an ideal time to develop long-range strategies concerning OPEC's role in the 1980s. In May, the Saudis invited Venezuela, Iran, Iraq, Algeria, and Kuwait to Taif, Saudi Arabia. The purpose of this gathering was informal discussion of the future course of oil prices and OPEC's role in minimizing the effect of unpredictable transformations in the oil market on the economies of OPEC members. A scenario developed by the Saudis captured the interest of all participants save Algeria. For the coming decades, this scenario envisaged the oil markets going through three cyclical periods: the surplus period (as it prevailed at the time of the Taif meeting) was to taper into a period of balance, which was to be followed by a critical period of short supply. Before the emergence of a dangerous shortage, as perceived by the Saudis, prices were expected to shoot up, thereby creating an economic slump and bringing the oil market back into balance again. Under these conditions, according to the Saudis, considerations of supply and demand, not OPEC, would be responsible for determining the prices of crudes in the future. It was therefore vital for OPEC to draw up a long-term strategy for the 1980s to secure its own role as a price-setting entity.[22] The Taif meeting led to the formation of a committee of OPEC ministers who were to study this issue and submit recommendations to the OPEC conference. The Iranian crisis of late 1978 dealt a severe setback to further consideration of this matter, however. It was not until 1981, with the reemergence of a soft market, that the oil states rejuvenated their interest in long-range planning with a view to safeguarding the longevity of OPEC.

The decline of the American dollar was the second issue OPEC needed to rectify. Although the American economy, after showing a deficit of $2.3 billion in 1974, recorded a substantial surplus of $11.6 billion in 1975, this surplus was reduced to $4.3 billion in 1976. Moreover, in 1977 the U.S. economy recorded a deficit of $15.3 billion, with forecasts for 1978 estimating a deficit of about $18 billion.[23] These deficits resulted in the steady decline of the dollar in 1977 and 1978. The majority of OPEC members favored an urgent dollar price adjustment. The OPEC conference in mid-June established a committee of OPEC experts under the chairmanship of the oil minister of Kuwait, Sheikh Ali Khalifah al-Sabah. This committee recommended that, starting from the fourth quarter of 1978, OPEC periodically calculate the price of Arabian Light marker crude on the basis of a formula (contained in the Geneva II formula of 1973) reflecting the movement of a basket of eleven currencies against the American dollar. This recommendation lacked majority support, however. The Saudi government was at first sympathetic to suggestions for adjustment of the dollar price quotations, but later showed little enthusiasm for quick incorporation of the corrective measures recommended by the OPEC experts. The decision-makers in Riyadh believed that any precipitous action aimed at tinkering with the value of the dollar, if it backfired, would be most costly for Saudi Arabia, the bulk of whose enormous foreign assets were in dollar denominations. Then, toward the end of August, Crown Prince Fahd, by rejecting the proposal to substitute a basket of currencies for the dollar, put an end to a variety of speculations. Even though Sheikh Ali later stated that the currency basket issue was still alive, this subject did not need further consideration because of the drastically altered oil market in the wake of the political revolution in Iran.

The Iranian Oil Shortfall

Toward the end of the third quarter of 1978, almost all bets were off that world energy markets would soon be facing a deluge, not a drought. Such a perception was based on a variety of supply-related facts. Alaskan oil was backing up on the West Coast of the United States and was being shipped over 7,000 miles through the Panama Canal to refineries in the Caribbean. Mexican oil had to seek markets as far afield as Eastern Europe. The North Sea was driving indigenous coal and African oil out of parts of the European market.[24] It was widely believed that a slowing in the growth of world oil demand and the anticipated rapid increase in non-OPEC oil sources would mean that OPEC production, which peaked in 1977, was likely to decline in the future. OPEC was expected to be most vulnerable to consumer pressure in the days ahead, since a number of populous OPEC nations were likely to be hard pressed to expand production at a time when world demand for OPEC oil was gradually declining. Based on projections of increased crude production of about 5 million b/d in the North Sea and about 2 million b/d in the Alaskan North Slope by 1982, predictions were that OPEC members would be forced to cut production so drastically that it would hamper the continued implementation of their economic development programs. This inability to produce the desired quantities of crude oil was expected to give birth to natural frictions within OPEC.[25] But this and similar forecasts[26] could not take into consideration the supply problems that were to grow out of the political revolution in Iran.

During the third quarter of 1978, oil production in Iran reached a peak of 6 million b/d. On October 31, 40,000 Iranian petroleum workers went on strike, a move that drastically reduced the crude output in that country. In the first week of November, Iranian crude oil production was reported to be around 1.8 million b/d, and total export shipments to be around 1 million b/d. The sudden loss of 5 million b/d was eventually expected to be recouped from increased production in other countries of the Persian Gulf. Saudi Arabia made a noticeable concession by allowing Aramco to raise production 9-10 million b/d. Despite this Saudi measure, the impact of the Iranian shortfall was immediately felt in the spot market, where prices escalated instantly to about 10-16 percent above the official OPEC prices. The major losers in the Iranian shortfall were the three largest suppliers of Iranian crude, the National Iranian Oil Company, British Petroleum, and Shell. All of these suppliers served *force majeure* notices to

Table 27. Initial Phases of the Iranian Oil Shortfall, November 1978 (1,000 b/d)

Date	Total Production	Export	Internal Consumption
Nov. 11	1,990	1,530	460
Nov. 12	2,115	1,651	464
Nov. 13	2,591	2,125	466
Nov. 14	2,968	2,443	525
Nov. 15	2,910	2,300	610
Nov. 16	3,064	2,386	678
Nov. 17	3,210	2,565	645
Nov. 18	3,345	2,678	667
Nov. 19	3,452	2,795	657
Nov. 20	3,734	3,102	632
Nov. 21	3,968	3,286	682
Nov. 22	4,682	3,975	707
Nov. 23	4,876	4,191	685
Nov. 24[a]	5,080	4,400	680

Source: *Middle East Economic Survey*, Nov. 20 and 27, 1978.

[a]Estimate.

their customers. BP's cutbacks from its Iranian supplies was reported to be around 40-45 percent, plus a 25 percent reduction on deliveries of other crudes; Shell introduced an across-the-board cutback of 11 percent. The American companies, such as Exxon, Texaco, and Mobil, without serving notices of *force majeure*, warned their customers of impending cutbacks.[27] In contrast, companies with the appropriate availability of crude oil were reaping a bonanza. At Rotterdam, the prices of premium gasoline rose by 43 percent, gas oil by 40 percent, and high-sulphur fuel by 16 percent over the second quarter of 1978.

The military government in Iran, which was created on November 6 to quell mounting anti-Shah rioting, issued threats of severe retaliation against the striking oil workers. On November 13, the workers responded by returning to work, and Iranian crude production began to record impressive gains (see Table 27) until November 24, when output from Khuzestan oil fields were expected to top 5 million b/d for the first time since mid-October. But in December the political situation in Iran took a turn for the worst. In response to a call by the exiled political leader Ayatollah Rouhollah Khomeini, 70 percent of the Iranian oil workers stayed off the job. This move crippled the Iranian oil sector, which ceased production on December 26.

The loss of Iranian crude to the world market underscored the acute dependence on OPEC of the Western industrial countries. Table 28 shows the

Table 28. OECD Dependence on Iranian, Arab, and Non-Arab Crudes, 1978

	Producer					
	Iran		Major Arab[a]		Major Non-Arab[b]	
Consumer	1,000 b/d	%	1,000 b/d	%	1,000 b/d	%
United States	525	8.0	3,052	46.5	2,679	40.8
Japan	807	17.9	2,431	53.9	1,403	31.1
Canada	114	17.2	192	29.0	388	58.5
U.K.	185	14.0	801	60.5	300	22.7
W. Germany	347	18.1	1,006	52.4	629	32.7
Italy	290	14.5	1,335	67.6	463	23.2
France	222	9.5	1,614	69.1	476	20.4
Netherlands	286	24.6	601	51.7	490	42.2
Western Europe[c]	1,658	16.3	6,311	62.2	2,766	27.3

Source: EIA, Department of Energy.

[a]Saudi Arabia, Kuwait, Libya, Iraq, UAE, and Algeria.

[b]Iran, Venezuela, Indonesia, Canada, Nigeria, and the Soviet Union.

[c]U.K., West Germany, France, Italy, the Netherlands, Belgium, and Spain.

dependence of selected OECD countries on Iranian, Arab, and non-Arab oil sources. The United States and the Netherlands were at two extremes in terms of their dependence on crude supplies from Iran. Despite varied degrees of dependence on Iranian crude, stocks in OECD countries were reported to be adequate to cope with about seventy days of supply interruptions. These countries were drawing down their inventories during the winter and were expected to feel the real effects of the Iranian shortfall in the second quarter of 1979, at the time of spring build-up (see Figure 4).

U.S. Energy Secretary James Schlesinger described the Iranian shortfall as "serious but not critical." The seriousness of this shortfall was reflected in the fact that the International Energy Agency (IEA), the Paris-based organization of the OECD nations, not only enhanced its close monitoring of the oil market but also initiated several measures under the International Energy Program.[28] (1) All oil companies based in OECD countries, which were required by law to participate in IEP, were asked to submit detailed monthly reports of their production, imports, exports, and stock changes. (2) Even though the IEA was not expected to trigger an emergency oil-sharing program, it was reported to be fully geared up to put this arrangement into operation under a continually deteriorating supply situation. (3) Under a sustained oil shortfall, the governing board of the IEA was also expected to consider potentially coordinated responses by member governments, which were designed to reduce consumption,

Figure 4: Outlook for Non-Communist Oil Supply/Demand, 1978-1980

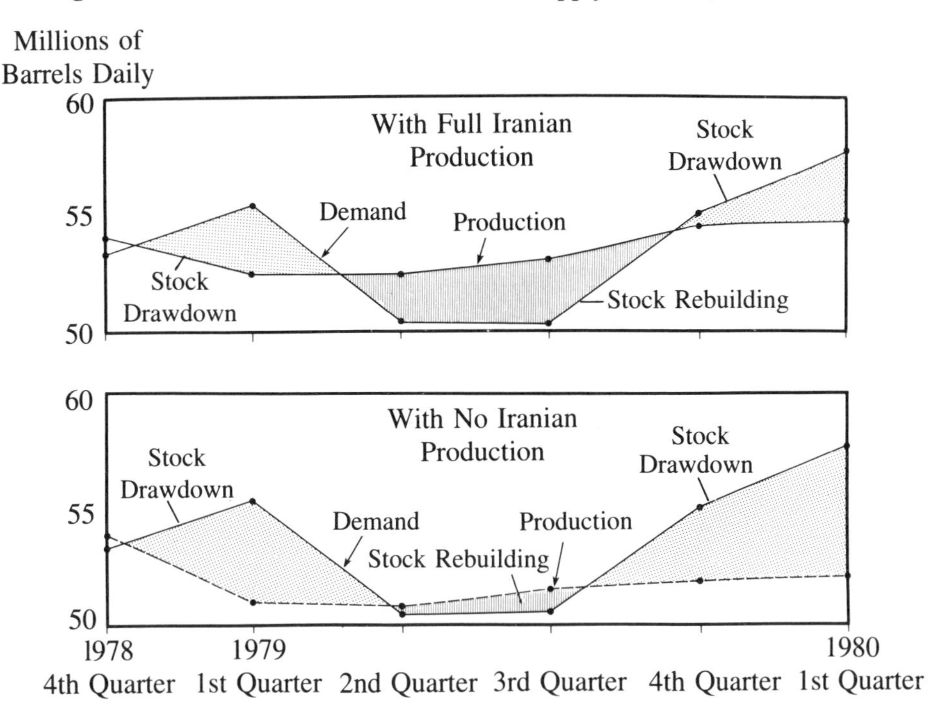

Source: U.S. Congress, Senate, Committee on Energy and Natural Resources, Hearing, *Iran and World Oil Supply*, 96 Cong., 1 sess. (Washington, D.C.: GPO, 1979), pt. 1, p. 18.

increase production, encourage fuel-switching, and insure equitable allocation of available production.[29]

While the OECD countries were scrambling to safeguard their supply situation, the oil-producing nations were busy hammering out their own responses to the Iranian shortfall, immediately preceding the OPEC conference in Abu Dhabi. Prior to the Iranian crisis, Saudi Arabia was producing at near capacity and had hoped to contain the OPEC price increase within the range of 5-10 percent. The emergence of a tight market now assured a price increase closer to 15 percent. After a series of pre-conference OPEC-wide consultations, an agreement was reached to raise prices in quarterly installments. The OPEC conference in Abu Dhabi, December 16-17, 1978, was used to hammer out the following schedule of price escalations: On January 1, the price of Arabian Light marker crude was to go to $13.335/b (a 5 percent increase); on April 1, it would be raised to $13.843/b (an increase of about 3.8 percent); on July 1, the

Table 29. Comparison of Spot and Official Prices, 1978-1981 ($/barrel)

	Gulf - Arabian Light (34°)					Libyan Zuetina (41°)		
	Spot	Off. (1)	Off. (2)	Diff. (1)	Diff. (2)	Spot	Off.	Diff.
1978								
Jan.	12.65	12.70	12.70	−0.05	−0.05	13.85	14.05	−0.20
Feb.	12.65	12.70	12.70	−0.05	−0.05	13.85	14.05	−0.20
March	12.65	12.70	12.70	−0.05	−0.05	13.75	14.05	−0.30
April	12.67	12.70	12.70	−0.03	−0.03	13.75	13.90	−0.15
May	12.72	12.70	12.70	+0.02	+0.02	13.75	13.90	−0.15
June	12.72	12.70	12.70	+0.02	+0.02	13.75	13.90	−0.15
July	12.77	12.70	12.70	+0.07	+0.07	13.75	13.90	−0.15
Aug.	12.79	12.70	12.70	+0.09	+0.09	13.85	13.90	−0.05
Sept.	12.80	12.70	12.70	+0.10	+0.10	14.00	13.90	+0.10
Oct.	13.00	12.70	12.70	+0.30	+0.30	14.50	13.90	+0.60
Nov.	14.90	12.70	12.70	+2.20	+2.20	16.25	13.90	+2.35
Dec.	15.00	12.70	12.70	+2.30	+2.30	16.75	13.90	+2.85
1979								
Jan.	17.50	13.40	13.40	+4.10	+4.10	19.75	14.74	+5.01
Feb.	23.00	13.40	13.40	+9.60	+9.60	26.00	15.42	+10.58
March	21.00	13.40	13.40	+7.60	+7.60	24.00	16.12	+7.88
April	21.50	14.55	16.35	+6.95	+5.15	24.50	18.30	+6.20
May	34.50	14.55	16.95	+19.95	+17.55	36.00	21.31	+14.69
June	34.00	18.00	18.00	+16.00	+16.00	36.50	21.31	+15.19
July	32.00	18.00	20.00	+14.00	+12.00	36.00	23.50	+12.50
Aug.	34.00	18.00	20.00	+16.00	+14.00	36.00	23.50	+12.50
Sept.	35.00	18.00	20.00	+17.00	+15.00	37.00	23.50	+13.50
Oct.	38.00	18.00	22.00	+20.00	+16.00	40.50	26.27	+14.23
Nov.	40.00	24.00	26.00	+16.00	+14.00	43.00	26.27	+16.73
Dec.	39.00	24.00	26.00	+15.00	+13.00	41.50	30.00	+11.50
1980								
Jan.	38.00	26.00	28.00	+12.00	+10.00	41.00	34.72	+6.28
Feb.	36.00	26.00	28.00	+10.00	+8.00	38.50	34.72	+3.78
March	36.00	26.00	28.00	+10.00	+8.00	38.00	34.72	+3.28
April	35.00	28.00	28.00	+7.00	+7.00	37.50	34.72	+2.78
May	35.50	28.00	30.00	+7.50	+5.50	38.50	36.72	+1.78
June	36.00	28.00	30.00	+8.00	+6.00	37.50	36.72	+0.78

Table 29. (continued)

	Gulf - Arabian Light (34°)					Libyan Zuetina (41°)		
	Spot	Off. (1)	Off. (2)	Diff. (1)	Diff. (2)	Spot	Off.	Diff.
1980 (continued)								
July	34.50	28.00	32.00	+6.50	+2.50	36.50	37.00	−0.50
Aug.	32.00	30.00	32.00	+2.00	—	33.50	37.00	−3.50
Sept.	32.25	30.00	32.00	+2.25	+0.25	33.50	37.00	−3.50
Oct.	37.50	30.00	32.00	+7.50	+5.50	38.25	37.00	+1.25
Nov.	41.25	32.00	32.00	+9.25	+9.25	42.25	37.00	+5.25
Dec.	39.25	32.00	32.00	+7.25	+7.25	39.50	37.00	+2.50
1981								
2 Jan.	39.50	32.00	36.00	+7.50	+3.50	42.00	41.00	+1.00
9 Jan.	39.25	32.00	36.00	+7.25	+3.25	40.50	41.00	−0.50
16 Jan.	39.25	32.00	36.00	+7.25	+3.25	40.25	41.00	−0.75
23 Jan.	38.75	32.00	36.00	+6.75	+2.75	40.00	41.00	−1.00
30 Jan.	38.00	32.00	36.00	+6.00	+2.00	39.50	41.00	−1.50
6 Feb.	37.50	32.00	36.00	+5.50	+1.50	38.50	41.00	−2.50
13 Feb.	36.50	32.00	36.00	+4.50	+0.50	38.25	41.00	−2.75
20 Feb.	36.50	32.00	36.00	+4.50	+0.50	38.25	41.00	−2.75
27 Feb.	37.00	32.00	36.00	+5.00	+1.00	38.25	41.00	−2.75
6 Mar.	37.00	32.00	36.00	+5.00	+1.00	38.25	41.00	−2.75
13 Mar.	36.75	32.00	36.00	+4.75	+0.75	38.25	41.00	−2.75
20 Mar.	36.50	32.00	36.00	+4.50	+0.50	38.25	41.00	−2.75
27 Mar.	36.50	32.00	36.00	+4.50	+0.50	38.25	41.00	−2.75
3 Apr.	36.50	32.00	36.00	+4.50	+0.50	38.25	41.00	−2.75
10 Apr.	36.50	32.00	36.00	+4.50	+0.50	37.75	41.00	−3.25
17 Apr.	35.75	32.00	36.00	+3.75	−0.25	37.50	41.00	−3.50
24 Apr.	35.50	32.00	36.00	+3.50	−0.50	37.25	41.00	−3.75

Source: *Middle East Economic Survey*, April 27, 1981.

Off. (1): Official sale price set by Saudi Arabia for Arabian Light marker crude.

Off. (2): Theoretical official price for marker crude used by other Gulf producers.

price was to be further raised to $14.161/b (a 2.3 percent increase); and on October 1, it was to reach $14.542/b (an increase of about 2.7 percent).[30]

Not long after the OPEC price schedule was announced, it became apparent that it would not hold under an exceptionally buoyant market. Meanwhile, spot prices were skyrocketing, largely for two reasons: the scramble for additional supplies by most OECD countries, oil companies, small refiners, and trading companies in the United States, whose supplies were trimmed by the Iranian shortfall; and rampant speculative stockpiling. In January 1979, Saudi Arabia temporarily raised its production ceiling for the first quarter from 8.5 million to 9.5 million b/d. The Saudis also stated that since ordinarily this extra 1 million b/d would have been produced later in the year, when a higher OPEC schedule would have been in force, the foreign participants of Aramco would be charged the fourth-quarter prices for supplies in excess of their normal entitlements.

Meantime, the increasingly ominous nature of the world oil market for consumers was becoming evident in the spot markets of Italy, Rotterdam, and certain Gulf areas. As is apparent from Table 29, in the fourth quarter of 1978, spot prices of Arabian Light marker crude remained at $2.30/b, and spot prices of Libyan Zuetina were $2.85/b, above their official prices. During the first quarter of 1979, spot prices of Arabian Light and Libyan Zuetina jumped to $9.60/b and 10.58/b, respectively, above official prices. Under such mounting uncertainties, it was only a matter of time before an outburst of unruly behavior on the part of the oil states would create increasing upward pressure on prices. In January, Abu Dhabi and Qatar imposed market premiums of about $1/b. In February, Libya and Kuwait followed suit, and Venezuela adopted a similar measure in March. In the same month, OPEC endorsed these actions of its members.

Even though the volume of crude involved in the spot market was very small,[31] the oil states were becoming impatient concerning reports of oil companies reaping a bonanza from the Iranian oil shortfall. Some OPEC members expressed their irritation at what they perceived to be gross double standards on the part of those industrial nations that were critical of the OPEC price increase under a tightening market but that at the same time allowed oil companies to make inordinate profits in downstream operations. As a manifest expression of their desire to share these profits, Libya, Algeria, and Nigeria called for an Extraordinary Meeting to reconsider the price schedules agreed upon during the last OPEC meeting. The Extraordinary Meeting in Geneva on March 26-27, 1979,[32] was marked by the now-familiar positions taken by the price hawks and the moderates. The hawks envisaged the tight supply situation as a golden opportunity to restore the eroded purchasing power of their revenues and to reverse in real terms the decline in prices of crude oil. There was a vain hope that Saudi Arabia, whose surplus production capacity was no longer a deterrent to the tide of price escalations, might be able to moderate some extreme pricing ideas put forth by the hawks. The end result was an "agreement-to-differ" compromise, a euphemism for a multi-tier price structure effective April 1. The first tier was comprised of spot prices. Prices of the North African Light low-sulphur crudes were included in the second tier. All Gulf states, save Saudi Arabia, added a surcharge premium, which was temporarily fixed at a flat rate of $1.20/b, regardless of quality,[33] and the third tier included these surcharged Gulf crudes. The unsurcharged Gulf crude sold by the Saudis formed the fourth tier. If the oil states intended to cut into the profit margin of the multinational oil corporations, this multi-tiered price structure created a situation quite to the contrary. The Geneva decision enabled the Aramco partners to acquire crude oil at a cost substantially lower than the rest of the oil industry could. It was also apparent that unless there was a significant upsurge in the production rate of non-OPEC oil, coupled with further upgrading of conservation techniques in the industrial countries, the price hawks would succeed in further escalating crude oil prices in the June and December meetings of OPEC.

The March 27 decision at Geneva was destined to meet the fate of the decision taken at the Abu Dhabi meeting under a sustained buoyant market. From 1975 through 1978, when world prices remained stable and spot prices were somewhat below contract prices, the spot market may have helped to keep down oil prices. The oil producers and consumers, not the spot market, determined the nature of supply and demand. In 1979, however, it was widely believed that the spot market was creating an illusion of scarcity, artificially inflating demand, and "ratcheting" contract prices to higher levels than they would have reached otherwise. Spot prices also affected the behavior of oil-exporting and -importing countries in three ways. First, the oil states frequently cited high spot prices as justification for raising contract prices and modifying contract terms. Second, high spot prices and expectations of future contract oil price escalations served as buyer incentives for unusually large purchases. Third, these large purchases and record high inventories were to play a crucial role in driving down oil prices from 1981 through 1983.[34]

In the second quarter of 1979, not only did a number of OPEC members, one after another, impose price surcharges, but also a great many were suspected to be withdrawing increasing amounts of crude from the contract sector of the market and feeding it into spot sales. Saudi Arabia was a rare exception in both instances. In fact, it was the only country that did not impose surcharges on medium and heavy crudes. The revolutionary Iran, on the other hand, was emerging as one of the most hawkish members on the issue of crude oil prices. Iran's hawkishness was manifested in its eagerness to impose surcharges, and it was reported to be one of the most active nations in terms of rerouting its contract crude into spot sales.[35]

The spot market remained at its peak despite the fact that the Iranian production, which remained at 2.5 million b/d in March (of which about 700,000 b/d went toward domestic consumption), reached the 4 million b/d mark in April. Part of the reason for the hyperactivity of the spot market was the Saudi move to rescind the January 1979 decision to raise production temporarily from 8.5 to 9.5 million b/d because of the availability of Iranian crude. In addition, the industrial consuming nations, in their scramble to rebuild oil stocks, plunged into a buying spree in an already tight market. The cumulative result was quantum spot price leaps in May and June, as shown in Table 29. Amidst this chaos, there was some murmuring from the moderates for a unified price structure. But for all producers except Saudi Arabia, which decided to forgo the option of playing the market for all it was worth, the multi-tiered price structure was responsible for reaping a bonanza. For this reason alone, the price hawks were expected to reject categorically any suggestions to reunify prices. In addition, the radically divergent views held by the Saudis and the hawks made reunification virtually impossible in the immediate future. The Saudis were reportedly pressing for a price level of $17/b for marker crude, while the price hawks were reluctant to accept any level much under $20/b. Unbeknownst to the price hawks, in allowing the spot market to set the prices of crude oil, OPEC was indeed abandoning its role as a price-fixing cartel.

As the OPEC states approached their June 27-28, 1979, conference in Geneva, it was abundantly clear that, thanks to the mounting supply- and price-related uncertainties of the world oil market, the pendulum of advantage was clearly swinging in favor of the price hawks. The Geneva meeting amounted to just another endeavor to ameliorate the chaotic conditions created by the multi-tiered price structure, surcharge leapfrogging, and other irregular practices. What emerged was a two-tiered price structure which appeared to be an orderly arrangement only in comparison to the multi-tiered price structure that it replaced. This feature, however, may have been the only positive feature of the two-tiered system. The first tier was comprised of Arabian Light at $18/b; the second tier included Iranian Light at $21-22/b, with a ceiling of $23.50/b, above which the official price of crude could not rise. The greatest weakness of this compromise was that it left differentials between individual crudes in a state of flux. Its chief strength, for what it was worth, was that OPEC members consented to keep the agreement in operation until the end of 1979.[36]

Saudi Oil Minister Yamani was not very happy with the general outcome of this meeting. As a remedial measure, he was counting on a shortfall reduction of about 2.3 million b/d through a combination of conservation policies by the Western industrial consumers (who were holding a summit meeting in Tokyo at around the same time) and a temporary escalation of production by his own country. Any increase in Saudi oil output was not expected to be viewed positively by the price hawks, whose tacit understanding at the Geneva meeting was that no member state would substantially raise its production level. In a change of heart, however, Saudi Arabia immediately announced that it would raise production temporarily to 9.5 million b/d. Even though the stated rationale for the increase was to generate the necessary funds for financing development projects, it was widely known that the real reason was to calm the market and to stabilize and unify prices. Within weeks after the Saudi violation of this tacit understanding, Qatar also breached the $23.50/b ceiling and the curbs on spot market sales agreed upon at the Geneva meeting. That country offered 3 million barrels of crude for sale at auction, for which it wanted $37/b, but which it was able to sell for only $34-35/b. In the three months following this transaction, the list of violators of the Geneva agreement of the past June continued to grow as Nigeria, Kuwait, Iran, Libya, Iraq, and Algeria raised their prices above $23.50/b.[37]

Toward the middle of November, Saudi Arabia was forced to reevaluate its oil policy with a steely sense of realism. In the past, the Saudis had threatened to use their production capacity and vast reserves to discipline what they perceived to be precipitous behavior by OPEC price hawks. In 1979 and under a hyperactive spot market, Saudi Arabia raised production rates during the first and third quarters (see Table A-5) and refused to enforce surcharges on its medium and heavy crudes, hoping to calm the spot market and to reunify prices. Such policies failed to produce the intended results, however, and Saudi Arabia remained frustrated by intermittent price escalations, leapfrogging of surcharges, and skyrocketing of spot prices. The Saudi decisionmakers were reported to be especially incensed by persistent reports that the American partners of Aramco—Exxon, Texaco, Socal, and Mobil—were buying Saudi Arabian Light at $18/b ($2-4/b less than comparable oil from other OPEC countries) but were charging about the same price for petroleum products made from Saudi oil as were their competitors, who were using higher-priced oil in their products.[38] Toward the end of the third quarter, the Saudi government was also disturbed by reports of huge profit margins made by the international oil companies. It was reported that Exxon, the largest of Aramco's foreign partners, earned $1.1 billion, a jump of 118 percent over the preceding year. That gain appeared "puny," however, when compared to the $993 million profit from its overseas transactions, a jump of 145 percent.[39] Aramco's other partners reported the following gains toward the end of 1979: Texaco, $1.76 billion (106 percent increase from net earnings of 1978); Mobil, $2.01 billion (78 percent increase from 1978); and Socal, $1.78 billion (64 percent increase from 1978).[40] In view of the preceding, Saudi Arabia, convinced that the chief beneficiaries of its lower prices and higher production were not the consumers of industrial nations but the treasuries of Aramco partners,[41] was expected to raise prices even before the next scheduled OPEC conference in Caracas. It should be pointed out that the Saudis were expecting a soft market to emerge in early 1980 and hoped that such a market would provide a suitable climate for the return of orderly behavior and for the reunification of OPEC prices. Oil affairs were destined, however, once again to defy rational calculations. On November

14, the ongoing conflict between Iran and the United States, whose embassy had been seized and diplomatic personnel held hostage by Iranian militants in Teheran, worsened when the United States froze Iranian assets in American banks and Iran suspended oil exports to the United States. Although the suspension of Iranian exports freed about 600,000 b/d of crude, at about the same time numerous OPEC members—Abu Dhabi, Iraq, Kuwait, and Venezuela—either slashed their production rates or were about to do so.[42] Such measures were not likely to calm the spot market. In such a climate, even the endeavors of the Saudis to reunify OPEC prices by raising their own was unlikely to entice the price hawks into forgoing additional increases in the prices of their own crude oil.

Four days before the Caracas meeting, as a culmination of their pre-conference consultations, Saudi Arabia, Venezuela, the UAE, and Qatar agreed to unify the prices of their crudes at $24/b. At the Caracas conference, December 17-20, 1979, Indonesia, Kuwait, and Iraq also realigned their prices at $24/b. By agreeing to raise the price of Arabian Light by $6/b (from $18 to $24/b), Saudi Arabia was painstakingly striving not only to defuse the price issue and to preempt attempts by more hawkish members to propel official prices toward spot market levels, but also ultimately to reunify OPEC prices. The Caracas conference, however, produced another now-familiar "agreement-to-disagree" outcome, and price reunification fell prey to the sustained tightness in the world oil market. While the price moderates stayed with their $24/b tier, the African producers (Algeria, Libya, and Nigeria) announced that they would raise the prices of their top-quality crudes to $30/b. As a further tribute to the chaos and uncertainties of the world oil market (which were expected to continue in 1980), the moderates raised their prices to $26/b within one week after the Caracas conference, and there were reports that the price of Saudi Arabian Light crude would also be upwardly readjusted to $26/b.

A Continued Drift

In the second half of 1979, various gloomy predictions concerning a potential economic slump in the coming year were beginning to set the tempo for 1980. The IMF was forecasting a period of severe strains for the world economy, marked by sluggish growth, persistent inflation, high unemployment, and poor policy options for the consuming nations at large. Moreover, the decline of output growth in the industrial world was predicted to be much sharper than had been generally expected. The OECD projected that higher-than-expected oil prices would mean that real (i.e., inflation-adjusted) economic growth in the industrial world would average only 2 percent rather than 2¾ percent in the second half of 1979 and the first half of 1980.[43] The mid-1980 projections grew progressively worse. It was reported that the price escalations since the end of 1978 had increased the net oil import bill of the OECD countries by an amount equivalent to about 2 percent of their GNP. Based upon this forecast, the real GNP of these countries was projected to be reduced by 5 percent by the end of 1981. It was also forecast that productivity was likely to decline through the remainder of 1980 and that the OECD area deficit on external current accounts were likely to be nearly $100 billion annually in the early part of 1980 but to decline to below $50 billion by the first half of 1981.[44] Forecasts for the second half of 1980 were even grimmer. The OECD countries flatly announced that their economies had entered recession.[45]

In the first quarter of 1980, industrial consumers were not buying as much for stockpiles as they had earlier in the winter. According to figures released by Exxon, the non-Communist world had entered a relatively mild winter with inventories 10-11 percent higher than normal. The IEA estimated that crude stock among its members totaled a record 3.066 billion barrels in January 1980 as compared with 2.81 billion barrels at the same time the preceding year.[46] The immediate impact of the slower pace of buyers' activities becomes apparent when one compares spot prices of Arabian Light and Zuetina crudes (see Table 29) in the fourth quarter of 1979 and the first quarter of 1980. Clearly, spot prices were heading downward. It was quite evident that the prolonged rowdiness of the spot market, coupled with intermittent crude price increases, was creating two ominous realities for OPEC. First, because of the sustained uncertainties of the Iranian oil shortfall, the Western industrial nations had perhaps overly insulated themselves from repetition of a similar happenstance. Second, and a necessary extension of the preceding, the financial resources thus exhausted by most OECD nations in building up huge oil stockpiles meant that they would be excessively cautious and conservative about allowing an ambitious growth of energy demands in their economies. Exxon's projection on this issue was that the non-Communist energy demand for 1980 could drop as much as 1 million b/d.

The preceding forecasts of gloom and doom did not seem to affect the intensity and pervasiveness of the uncertainties of the sellers' market, however. Convincing evidence of this condition emerged in the decision of Saudi Arabia on January 28 to raise the price of Arabian Light crude from $24 to $26/b, effective from January 1. This decision was made to unify the Saudi price with those of other members—Venezuela, the UAE, Qatar, Kuwait, Iraq, and Indonesia. Prior to the Saudi action, three groupings within OPEC could be identified with specific reference to their prices: exporters who used the Saudi price of $24/b, price hawks who charged as much as $34.72/b; and a third group who, for political or geographical reasons, did not fit into the preceding two.[47] Within twenty-four hours of the Saudis' decision to escalate prices, their hopes of unifying crude oil prices were dashed. Kuwait, the UAE, Iraq, and Qatar, whose prices were aligned on the $26/b marker belatedly attained by Saudi

Table 30. OPEC Prices before and after Saudi Decision of January 28, 1980 ($/barrel)

Producer	January 28	February
Libya	34.72	38.72
Nigeria	34.48	38.69
Algeria	33.00	37.21[a]
Indonesia	27.50[b]	29.50[b]
Iran	30.00	32.50
Venezuela	28.75	30.75
UAE	27.56	29.60
Qatar	27.42	29.42
Kuwait	26.00	28.00
Iraq	25.96	28.00
Saudi Arabia	24.00	26.00

Sources: *Oil and Gas Journal,* Jan. 7, Feb. 4, 1980; *Middle East Economic Survey,* Feb. 4 and 11, 1980.

[a]Includes $3/b refundable exploration surcharge.

[b]*Oil and Gas Journal* reported the Jan. 28 price of Indonesian oil as $30.75; if one accepts that scale, the Indonesian price in Feb. becomes $32.75.

Arabia, raised their contract prices by $2 to $28/b and made them retroactive to January 1. Within a week there ensued another round of leapfrogging by Nigeria, Algeria, Libya, and Indonesia. Table 30 shows the state of price escalations within three weeks of the Saudi price increase.

In the second quarter of 1980, the spot market showed clear signs of weakness. An important question under the threat of a soft market was whether the oil states would be able to manage a possible mini-glut. Financially there was little doubt that OPEC members were in a comfortable position, thanks to the chaotic market conditions following the Iranian oil shortfall. OPEC's current balance accounts, after showing a deficit of $1 billion in 1978, skyrocketed to a surplus of $111 billion in 1980 (see Table A-8). In fact, some oil states were even expected to welcome production cutbacks, which, as previously noted, were announced but not implemented by many countries. In a potentially soft market, Saudi Arabia was expected not only to regain its leverage as a swing producer but also to succeed in reunifying OPEC prices by sustaining its production level at 9.5 million b/d through the second quarter.

The possibility of a soft market suffered a setback on April 7. President Carter, in addition to severing diplomatic relations with Iran, imposed an economic embargo on that country and secured support from Japan and Canada to tighten economic screws on Iran. As a retaliatory measure, Iran announced a series of energy and trade agreements with the Communist bloc nations. These actions were not expected to be conducive to an immediate softening of the market. So, instead of using its surplus production capacity to unify OPEC prices, Saudi Arabia relied on raising the price of Arabian Light crude from $26 to $28/b as an enticement for the price hawks to accept price reunification. Yet, despite this latest increase, there remained a huge gap between the price of Arabian Light and the highest priced Algerian crude, which was being sold at $38/b, including a $3 exploration premium (see Table 30). Given such wide divergences, the chances of price unification remained slim at best. In less than a week, Kuwait, Iraq, and Abu Dhabi initiated a new leapfrog by raising their prices from $28 to $30/b. The realities of the oil market were further reflected in the so-called "price compromise" worked out at OPEC's conference in Algiers on June 9-11. This arrangement established a theoretical marker ceiling of $32/b, which was $4/b higher than the price of Arabian Light. It was agreed also that differentials added over and above this ceiling were not to exceed $5/b. Even this compromise was destined to be set aside by the price hawks sooner rather than later.[48]

Saudi Arabia continued to produce at the rate of 9.5 million b/d, hoping that this high output, coupled with sluggish economic activity in the industrial consuming nations, would create downward pressure on spot prices. The economic exigencies of this period were such that the oil companies and industrial consuming countries, fearful of another political crisis that might create an oil shortfall, continued to build their inventories. The United States resumed filling its strategic petroleum reserves. The European countries and Japan were reported to be aggressively accumulating oil supplies, a major portion of which were obtained as a result of direct dealings with the governments of oil-producing countries. As a consequence, world oil inventories were reported to be at an unprecedented 5.3 billion barrels. This flurry of activity in the international oil market maintained an upward pressure on crude prices despite the fact that spot prices were falling.

In August, it appeared almost certain that petroleum prices were moving downward (see Table 29). The oil states began to reduce their premiums, and there were reports that some premiums were either eliminated or ignored. But at the end of a tri-ministerial meeting in Vienna on September 15-17, Saudi Arabia was persuaded by its OPEC partners to raise the price of Arabian Light from $28 to $30/b. A noteworthy aspect of this conference was the renewed interest of the oil states in long-term strategy for OPEC. The draft plan of this strategy, which was considered by the participants, proposed adoption of a floor pricing formula. This formula was designed to maintain the purchasing power of OPEC revenues through periodic adjustments for inflation and fluctuations in the value of currencies of ten major OECD countries, and to minimize the gap between the real price of oil and the cost of alternative substitutes. According to the draft plan, this formula was applicable only when the market was roughly in balance. Under tight market conditions, prices were to be allowed to rise above the floor level. When shortages disappeared, the oil states were to use discretion as to whether to freeze prices or to establish a new price floor at the higher level. A significant aspect of this draft plan was that it provided for production programing as a mechanism to preserve the price structure. The specifics of this production programing were to be negotiated by the participating states whenever OPEC decided to use this mechanism. The Saudi willingness to participate in production programing was contingent upon fulfillment of the following: (1) reunification of OPEC prices based on a single market crude, along with an agreed system of differentials and no provisions for market surcharges of discount allowances; (2) adoption of the proposed floor pricing formula by OPEC; and (3) implementation of production programing on the basis of a fair distribution of production cuts.[49]

Consumers' worst fears of a possible recurrence of an oil shortfall resulting from a political crisis in the Middle East came true with the outbreak of war between Iran and Iraq in late September 1980. The eruption of hostilities between these two major oil producers eliminated between 3.35 and 3.5 million b/d of crude (between 12 and 16 percent of OPEC exports and about 10 percent of total Western oil imports) from the world market.[50] The short-term effects of war on the industrial consumers were not expected to be too severe unless the Strait of Hormuz was blocked. The United States, which was importing only 30,000 b/d from Iraq at that time, was one of the most insulated countries, but certain other major consumers were not so well protected. Spain was expected to lose 30 percent of its import needs from Iran and Iraq; 20 percent of French, 10 percent of Japanese, and 10 percent of Italian needs were to go unmet because of the loss of Iraqi crude; and Brazil was expected to lose 40 percent of its import needs from Iraq and 10 percent from Iran.[51] East European countries such as Bulgaria, Czechoslovakia, East Germany, Hungary, Poland, and Romania were collectively importing about 200,000 b/d from Iraq and about 130,000 b/d from Iran.[52] Despite this shortfall, most industrial consumers were reported to have about 110 days' supplies on hand. The loss of supplies for the East European countries was expected to be met in the short run by the Soviet Union.

The fourth quarter of 1980 was a period of the most comfortable worldwide stock position ever recorded, but the world oil market did not appear immune to chaotic conditions emanating from the oil shortfall. This time, the problem was that the stocks were unevenly distributed, and industrial consumers were not equally insulated from a potential shortfall. Moreover, it was not clear whether those consuming nations with comfortable stocks would draw them down in order to share with those countries with low stocks.[53] In consequence, the latter countries were expected to turn to the spot market to build their inventories. The anticipated triggering of such activities reinvigorated spot prices in October and November, which had been declining since August. Toward the end of November, there were again dangerous signals that an upsurge in spot prices (regardless of volume) would trigger an upward spiral of the entire price structure. Overly concerned about the deleterious impact of spot prices on a relatively stable oil market, Saudi Arabia, Kuwait, the UAE, and Qatar decided to increase their rate of production.[54] It was unthinkable, though, that the price hawks would forgo this opportunity to press for additional price escalations at the next OPEC meeting.

The December 15 conference of oil states in Bali once again failed to reunify OPEC prices. As expected, Saudi Arabia raised the price of Arabian Light from $30 to $32/b. A new ceiling of $36/b was established as the base for applying a maximum differential of $5/b for premium crudes, thus making an overall ceiling of $41/b for all OPEC crudes. But few producers were expected to take full advantage of the new formula while the future of the oil market appeared cloudy for 1981 and beyond.

OPEC's Widening Window of Vulnerability

The 1980s appear to fit the scenario for the collapse of OPEC so frequently described by M.A. Adelman.[55] The oil states had become genuine sellers of oil. The demand for OPEC oil had been low because of a combination of the general economic downturn that had prevailed in the industrial consuming nations from 1981 through 1983, sustained practices of conservation and fuel-switching, and increasing competition from non-OPEC oil producers. The monopoly of the major oil companies, which were lifting 80 to 90 percent of OPEC exports, had been replaced by a high degree of competition. Each petroleum-exporting country had been transacting with twenty to forty customers, including previous concessionaires, American independents, oil companies from Japan, Europe, and the Third World countries, and others. The increased number of companies operating outside the old integrated channels of the multinational oil corporations had enhanced the significance of the spot market. In the previous decades the frequent mismatches between oil supplies and requirements were internally corrected by the horizontally and vertically integrated multinational oil corporations. These corporations relied on techniques such as inventory changes, redirection of tankers, and changes in the refinery input mix. Since 1979, however, oil buyers had been increasingly relying on the spot market to correct these imbalances,[56] an arrangement that had greatly diminished the power of OPEC. These factors had implications for the manueverability of OPEC.

The year 1981, like 1980, began with predictions of the emergence of a soft market, barring an unforeseen political crisis affecting either supplies or prices of crudes. Any scenario of a soft market automatically involved the reinstatement of Saudi Arabia as a swing producer, using its two-pronged strategy of production and pricing to bring about stability of prices.

It will be recalled that, during 1977 and on numerous occasions in 1979 and

1980, Saudi Arabia had unsuccessfully tried to reunify OPEC prices. During the first quarter of 1981, however, the chances of Saudi success appeared high for several reasons. First, the demand for crude in the OECD countries remained weak. Second, the combined crude production from Iraq and Iran was reported to be between 2 and 2.7 million b/d (see Table A-2). If continued through the second quarter, this increased production, coupled with an escalated Saudi production of 10.3 million b/d, was bound to create a surplus supply situation in the world market. Third, spot prices, after showing a short-lived upward movement in the fourth quarter of 1980, were least brisk for Arabian Light and African crude. In addition, the African producers, under a depressed spot market (see Table 29), were increasingly being pressured to lower or even ignore their differentials. In fact, in the second quarter of 1981, even the spot prices of mainstream Arabian Light crude (the official price of which was raised in January from $32 to $36/b by Kuwait, Iraq, and Qatar) dipped below the official price of $36/b.

The Saudi insistence on price reunification grew almost in direct proportion to the increasing softness of the world oil market. As it appeared certain that the surplus supply situation would prevail at least through 1981, the OPEC price hawks clamored for price and supply stability. The foremost expression of their concern was the sudden rejuvenation of their interest in the long-term strategy plan. This plan had temporarily lost its relevance in the wake of the Iran/Iraq war when the price hawks were more interested in exploiting the anticipated soft market in the fourth quarter of 1980. With the reemergence of a soft market, which also revived the traditional divisions between OPEC price moderates and hawks, the proposed plan became the focal point of dissension. The price hawks, especially Libya, were calling for a revision of the plan in order to strengthen the price formula that had been considered at the tri-ministerial conference in September 1980. Most important, the price hawks insisted on incorporating production programing. Saudi Arabia, however, consistent with its stated position at the same conference, wanted an OPEC-wide price unification as a precondition for its consideration of the long-term strategy plan. Moreover, as a symbol of its commitment to price reunification, Saudi Arabia sustained production at 10.3 million b/d, thus glutting the market, and reduced the price of its "war relief crude" from $36 to $34/b, effective from April.[57] At the next OPEC conference in Geneva, Saudi Arabia could not persuade the price hawks to lower their marker price of $36/b to meet the $34/b mark, nor could the price hawks persuade Saudi Arabia to raise its price. The result was an uneasy stalemate. A significant decision made at this conference was to cut output by some 10 percent, or roughly 1.3 million b/d, effective June 1. Saudi Arabia promptly dissociated itself from this decision.

As expected, the African producers bore the brunt of the Saudi-engineered glut. By August, Libyan production plummeted from 1.7 million to around

800,000 b/d and was expected to go down still further. Algerian production was estimated to have dropped from a high of 1.1 million to 800,000 b/d during the third quarter of 1981. Nigeria was reported to have absorbed an equally substantial loss in production.[58] In the meantime, the market remained in a slump, strengthening Saudi opposition to the $36/b price and the majority view that Saudi Arabia should make substantial production cutbacks. At their Geneva meeting in August, OPEC members again failed to unify prices. Saudi Arabia, as a gesture of goodwill, reduced production by 1 million b/d, but this action was not expected to reduce the prevailing surplus because the consuming nations and oil companies were expected to destock their inventories during the fourth quarter of 1981. The plight of the African producers was worsening, and it was only a matter of time before one of them would cross the threshold of continued resistance and start cutting prices. Nigeria turned out to be that country. Its production dropped from 2.1 million b/d at the beginning of the year to somewhere between 500,000 and 600,000 b/d in August. On August 26, Nigeria announced a $4/b discount on its $40/b oil, and Nigerian crude was reduced to $34/b with a differential of $2/b. The Nigerian decision created immediate pressure on Libya and Algeria. If these two countries resisted comparable price cuts, the boost in Nigerian sales would be at their expense. Both, however, refused to budge and maintained their prices at $40/b. The Gulf states, adhering to their scale of $36/b, also maintained a wait-and-see attitude.

The majority of OPEC members reached a breaking point toward the end of October. Until then, however, their behavior would have fitted a textbook description of promotion of national interests. For instance, it was not the traditional hawks (Iran, Libya, and Algeria), but Iraq and Venezuela that were responsible for undermining price reunification efforts at the August meeting. Iraq and Venezuela had very little to gain from reunification. Venezuela had not yet suffered from the oil glut. Iraq, on the other hand, was suffering financially from the Gulf war and did not anticipate extra sales by cutting prices. The multi-tiered price structure, nevertheless, was beginning to crack under the weight of market glut. Iraq was reportedly cutting prices by $4/b; Libya allowed a variety of tax and royalty exemptions to Occidental, which amounted to a cut of $4/b; and Kuwait had extended the credit period to its customers by an extra 30 days, a move which amounted to a $1/b discount on the official price of $35.50/b.

Toward the middle of September, it was reported that Iran, Iraq, Kuwait, and Qatar were starting to feel price pressure from their customers. By the middle of October, Venezuela, the last of the holdouts, expressed willingness to accept the $34/b scale as the basis for reunification. Toward the end of October, Saudi Arabia finally attained its long-cherished goal of subjugating the OPEC majority through its leverage as a swing producer. The continuance of a soft market throughout 1981 was the ideal circumstance for Saudi Arabia. At the conclusion of its October 29 meeting, OPEC announced the reinstatement of the

Table 31. OPEC Production, 1977-1983 (Total OPEC Production Capacity: 35 million b/d)

Year	Total OPEC Production	
	1,000 b/d	% of Capacity
1977	31,278	89.4
1978	28,805	82.3
1979	30,928	88.3
1980	26,890	76.8
1981	22,624	64.6
1982	18,660	53.0
1983	17,320	49.58

Sources: *Monthly Energy Review,* April 1983; *Oil and Gas Journal,* Jan. 31, 1983; March 11, 1985.

[a]*Middle East Economic Survey,* Sept. 29, 1980. For a country-by-country breakdown, see Table A-1.

old principle of pricing crude oil around the Arabian Light marker crude, which would also serve as the basis of price reunification. Under this arrangement, Saudi Arabia agreed to raise its price from $32 to $34/b, while remaining OPEC members lowered their prices from $36 to $34/b. The participants also agreed to freeze prices until the end of 1982 and not to allow differentials to exceed $4/b. Another important outcome of this conference was Saudi Arabia's decision to lower production and return to the old official ceiling of 8.5 million b/d.

To put OPEC's price increases in perspective, it should be noted that the average increase in world crude oil prices in 1979 was 109 percent (from $13.74 to $28.84/b). This average rose by another 23 percent by January 1, 1981 (to $35.49/b). Thus, during the two-year period, OPEC prices increased by 158 percent. As a result, certain realities, which were to prove quite painful for OPEC in the coming years, were beginning to be apparent in the international oil market in 1982. First and foremost among these was the expectation that imports of OPEC oil, which had played a significant role in demand in the non-Communist countries, would shrink significantly. One reason was the declining world demand for crude oil, some of which was due to structural changes, such as fuel-switching and conservation, and some of which was due to the continuing recession. As shown in Table 31, OPEC's total production dwindled from 30.9 million b/d in 1979 to 22.6 million b/d at the end of 1981, a shrinkage of about 65 percent. From the oil producers' perspective, the statistics appear more ominous. OPEC's share of the total non-Communist production was 68.6 percent in 1976 but took a nosedive to 59.5 percent in 1980 and to 54.4 percent in 1981. By the end of 1982, estimates indicated OPEC's share to have slipped to 48.45 percent.[59] Energy efficiency had improved every year since 1970, with energy consumption declining from 62,500 BTU/$ of real GNP in 1970 to an estimated 47,700 BTU/$ in 1982, a cumulative decline of 26.6 percent (also Table A-9). Oil energy consumed per dollar of real GNP fell again in 1982, for the fourth consecutive year. Since 1977, oil energy per GNP dollar had dropped from 27,060 BTU to 21,190 in 1981 and to a projected 20,430 BTU in 1982. Coal consumption was expected to increase by 2 percent in 1982, and nuclear generation of electricity by 5.1 percent. The lower level of economic activity, combined with the OPEC price escalation of previous years, was anticipated to produce a 1.2 percent decline in energy requirement in 1982 in addition to the 2.8 percent reduction in 1981. Energy consumption reached a peak of 78.91 quads in 1979, but between 1979 and 1982 higher energy prices caused a reduction of 7.6 percent in total energy consumption, while real GNP increased by 1.1 percent.[60]

Meanwhile, the Saudis were becoming concerned over the sustained slump. They let it be known that they would be open for persuasion if asked by the Persian Gulf allies to discuss the OPEC price structure. But instead of reducing its production level of 8.5 million b/d, which might have firmed up prices, Saudi Arabia maintained a public posture of relying on market forces to bring about the necessary adjustments. Toward the end of the first quarter of 1982, it became quite apparent that OPEC must not only think the unthinkable but also take action on it. The unthinkable was incorporation of production programing, to which many OPEC members had paid lip service in the past two decades but on which they had never been willing to act.

At the March meeting, OPEC, in the manner of a true cartel, agreed for the first time to implement production programing by establishing an OPEC-wide production ceiling of 17.5 million b/d. Consistent with its past objection to production programing, Saudi Arabia was not a party to this decision, but, in a separate decision that was clearly aimed at supporting the $34/b price level, the Saudis announced they would lower their production to 7 million b/d. Although all OPEC members agreed to abide by their production ceilings, the prospects of successful implementation in the near future remained cloudy, at best. Iran and Iraq, locked into a prolonged armed conflict, were already producing a little over 1 million b/d each. These countries needed additional funds not only to bankroll their war efforts but also to sustain their wartorn economies. Nigeria, whose production remained at 1.3 million b/d, was finding fewer customers at $36.50/b, especially at a time when Britain and Norway were offering their light crude at $32.50/b. The output figures shown in Table 32, when examined in light of the production capacity of each OPEC member (provided at the bottom of Table 31), indicates that the most potent force operating against production programing was that most OPEC countries were already producing at as little as 50 percent of capacity, while their capital needs were increasing rapidly because

Table 32. Comparison of OPEC Production, First Half of 1981 and of 1982 (1,000 b/d)

Country	First Half, 1981	First Half, 1982	% of Change
Algeria	867	800	−27.0
Ecuador	212	204	−8.0
Gabon	145	142	
Indonesia	1,620	1,367	−14.8
Iran	1,517	1,504	−14.3
Iraq	800	1,030	+35.0
Kuwait	1,107	633	−40.4
Libya	1,490	868	−53.0
Neutral Zone[a]	434	273	−32.0
Nigeria	1,695	1,317	−22.0
Qatar	465	341	−30.1
Saudi Arabia	9,966	7,123	−26.9
UAE	1,576	1,286	−24.0
Venezuela	2,169	1,674	−22.9
Total	24,063	18,562	−24.0

Source: *Oil and Gas Journal,* July 26, 1982; July 25, 1983.

[a]Shared by Kuwait and Saudi Arabia.

of international inflation and other forces. In addition, the combination of dwindling demand for crude and a continued stock drawdown by the international oil companies was expected to force the oil states to introduce further cutbacks in the coming months. Under these conditions, it was hard to imagine that the capital-short oil states (or high absorbers, listed in Table A-8) would not indulge themselves in competitive bidding, price discounts, and other unfair tactics in order to sell more, practices which in turn would violate the letter as well as the spirit of production programing. By July, internal frictions over differentials and over reports of violation of quotas by some oil states, which were idiosyncratic of a soft market, began to surface. Saudi Arabia insisted that the differential charges of the African crudes, which remained at $1.50/b, be raised to $3/b. The African oil producers, on the other hand, did not wish even to discuss such increases.

The violation of quotas was emerging, perhaps, as a major and sustained source of irritation between the price hawks, especially Iran, and the moderates within OPEC. To understand the problem of quota violations, one must understand the role of non-OPEC producers. To consolidate their position in the international oil market, these producers (primarily the U.K., Norway, Mexico, and to a lesser extent Egypt) operate in direct competition with OPEC. In a tight market, they sell their crude at a price higher than that officially set by OPEC. In a soft market, shifting their concern from prices to volume, these producers are readily amenable to price cuts in order to maintain their share of the market. Most OPEC members, on the other hand, have been more interested in maintaining their price structures, and, as a matter of general practice, have been willing to lower production in order to protect prices. As a result of production increases from non-OPEC sources, OPEC in the 1980s has emerged as a residual supplier, i.e., a supplier forced to absorb non-OPEC production increases by lowering its own prices in order to protect its price structure. Another source of pressure on OPEC in the 1980s stems from members such as Iran, Nigeria, Libya, and Venezuela. In the 1970s these countries behaved in the manner of non-OPEC volume-maximizers by shaving prices or offering other discounts to maintain their share of the market at the expense of other OPEC states. It should be pointed out, however, that Iran under Khomeini has emerged as the chief violator of production quotas, while Libya, Nigeria, and Venezuela have been less consistent violators.

In July 1982, the confluence of intra-OPEC frictions over differentials, violation of price quotas, and the price-cutting/volume-maximizing activities of non-OPEC producers brought about the collapse of the production programing arrangement. The increasing softness of the market created a four-tier price scale in the world market in which the Saudi Arabian OPEC marker at $34/b emerged as the unprecedented highest priced crude. The non-OPEC marker was being sold by the North Sea producers and Mexico at $31-32/b. Libya and Iran were reportedly selling their crude at $32 and $31/b, respectively. With Aramco production falling to 6.3 million b/d, Saudi Arabia began lobbying for adjustment for differentials. Between July and December, Iran, Libya, and Venezuela continued to overproduce. And toward the end of 1982, total OPEC production was estimated at 18.584 million b/d, while the production of the remaining non-Communist areas was reported at 19.768 million b/d. At this level, OPEC's production was off 17.6 percent (3.957 million b/d) from 1981, after falling 16.9 percent (4.54 million b/d) in 1981. A full 80 percent of the 1982 OPEC production loss (3.157 million b/d) was borne by Saudi Arabia.[61] A comparison of the rates of OPEC and non-OPEC production at the end of 1982 is provided in Table 33. In view of its situation, OPEC was faced with another unprecedented choice: that is, to consider reducing the price of its marker crude early in the coming year if world oil demand remained in a slump and if its member states failed to incorporate production programing.

In 1983, volume maximization by the North Sea producers, especially the U.K.; was one of the most significant sources of consternation for OPEC. In the aftermath of their failure to incorporate production programing in their December meeting, the oil states focused their attention on defending the $34/b price structure, which was under tremendous downward pressure. If OPEC's hopes for salvaging its price structure were based upon expectations of accelerated demand in OECD countries, the economic forecast for these countries was

Table 33. Comparison of OPEC and Selected Non-OPEC Production, 1982 (1,000 b/d)

OPEC Producer	1982 Production (12-month avg.)	% of Change from 1981
Algeria	775	−4.6
Ecuador	209	−0.9
Gabon	141	−6.6
Indonesia	1,344	−16.2
Iran	1,998	+50.0
Iraq	914	−0.3
Kuwait	667	−29.0
Libya	1,203	+9.0
Neutral Zone[a]	305	−17.8
Nigeria	1,289	−10.0
Qatar	331	−18.3
Saudi Arabia	6,364	−34.0
Venezuela	1,891	−10.3
Abu Dhabi	866	−24.3
Dubai	356	−0.8
Sharjah	7	−30.0
Total	18,660	−17.2
Non-OPEC Producer	**1982 Production (12-month avg.)**	**% of Change from 1981**
United States	8,671	+1.0
Canada	1,241	−0.9
Mexico	2,749	+1.2
U.K.	2,117	+1.2
Norway	520	+1.0
USSR	12,053	+1.0
Egypt	670	+1.1
Total	28,021	+1.0

Sources: OPEC data from *Oil and Gas Journal,* March 14, 1983, p. 26; data for U.S., Canada, Mexico, U.K., and USSR from *Monthly Energy Review,* June 1983, p. 99; data for Norway and Egypt from *International Energy Annual, 1983* (Nov. 1984), p. 16.

[a]Shared by Kuwait and Saudi Arabia.

anything but encouraging. The OECD governments responded to the oil price rise of 1979-1980 by implementing tight monetary and fiscal policies aimed at decelerating demand and by lowering the inflation rates between 1979 and 1982. The effects of these policies became apparent only in 1981 and 1982. The 1983 projections for Europe were for very weak GNP growth until mid-1984, when the output was expected to be only 1½ points higher than two years earlier. In the United States, on the other hand, a pickup in GNP growth seemed likely in 1983, and in Japan, given the tight fiscal stance planned for 1983, domestic demand was expected to grow more slowly than usual.[62] Oblivious to these forecasts, non-OPEC producers decided to follow up their volume-maximization with aggressive pricing and marketing policies, and the Gulf producers were convinced that, to keep their share of the market from dwindling further, they would be forced to reduce the price of OPEC marker crude by $4/b. In January 1983, Saudi Arabia, Kuwait, the UAE, Qatar, Nigeria, Indonesia, and Iraq reached an informal agreement to that effect.

Shortly thereafter, the non-OPEC factor complicated the matter. The British National Oil Corporation (BNOC) was reportedly under heavy pressure from its customers for a price reduction of at least $2.50 to $4/b. Such an action would have reduced the price of North Sea crude by $4/b, putting it $6/b below the OPEC benchmark, and would have forced the African producers, notably Nigeria and Libya, to lower their prices. In fact, toward the end of January, the effects of Britain's anticipated price cut emerged in the form of lower sales for Nigerian and Libyan crudes. Iran was expected to encounter similar difficulties in meeting a sale target of 2.5 to 3 million b/d. The expected BNOC price cut served as a death blow to the informal agreement reached by the seven OPEC members in January.

In February, BNOC and Norway announced reductions of $3/b and $3.50/b, respectively, in their contract sales effective the first of that month. OPEC's response was contingent on the outcome of a complicated compromise that was to emerge soon. Nigeria announced that it would match the drop in the price of North Sea crudes ($3.50/b) by reducing the price of its crude by $5.50/b (from $35.50 to $30/b), also retroactive to February 1. Nigeria, expressing its resolve to match cent for cent any further cuts by Britain and Norway, also stated that its price cut was negotiable provided the North Sea producers, its main competitors, were brought into consultations. Nigeria's decision to realign prices with the North Sea producers reflected a new dimension of the uncertainty faced by OPEC in the 1980s. In the Western Hemisphere, the price of Venezuelan crude was being similarly aligned with the price of Mexican crude and that of the U.S. market. And Indonesia, though still aligned with the OPEC marker, was expected to be swayed heavily by Japan and the activities of other non-OPEC Asian exporters. The Gulf producers (excluding Iran and to a lesser extent Libya), who had lately emerged as the chief guardians of OPEC, were

Table 34. OPEC Production Quotas, March 1983 (1,000 b/d)

Producer	1,000 b/d
Algeria	725
Ecuador	200
Gabon	150
Iran	2,400
Iraq	1,200
Indonesia	1,300
Kuwait	1,050
Libya	1,100
Nigeria	1,300
Qatar	300
UAE	1,100
Venezuela	1,675
Total	12,500

Source: *Petroleum Economist*, Dec. 1984, p. 444.

faced with many uneasy choices. At one extreme, they had the option of doing nothing, which would have led inevitably to the de facto, if not de jure, collapse of OPEC as Nigeria and Venezuela were left to gravitate to their respective regional markets. At the other extreme was an option whereby the Gulf producers would have triggered a price war by matching the North Sea (and, by extension, African) producers cent for cent. This option would also have led to the dismantlement of OPEC. The middle-of-the-road option was to seek collectively an optimal strategy which suited the economic objectives of all parties. This option was also welcomed by the North Sea producers and Mexico, since none of them welcomed a price war that they were more certain to lose. Thus the quest for an optimal strategy became the focal point of a flurry of activity by all parties in London.

The London meeting was yet another unprecedented step, a consultation between OPEC and some non-OPEC producers. After several marathon negotiating sessions, a fragile compromise emerged. According to this compromise, OPEC, for the first time in its history, reduced the official price of Arabian Light marker crude by $5/b, from $34 to $29/b. This gave the North Sea, Algerian, and Libyan crudes a differential of $1.50/b. In the case of Nigeria, OPEC was forced to make an exception by allowing a differential of $1/b above the Arabian Light marker crude. The London agreement also established for OPEC output an overall production ceiling of 17.5 million b/d for the remainder of 1983. The breakdown of quotas for individual OPEC members is shown in Table 54. Saudi Arabia agreed to act as a swing producer, making up the difference between production quotas and 17.5 million b/d or total demand for OPEC oil. Undoubtedly the burden of this agreement was borne by Saudi Arabia. The member states gave the usual assurances of observing the production quotas and of avoiding offering discounts of any kind. But it was apparent that the willingness of a majority of them to stand by these assurances was entirely dependent on the future shape of the world oil market. The survival of this agreement was also contingent upon too many "ifs" over which OPEC had no control. For instance, this agreement was expected to last only under the following conditions:

(1) If the North Sea producers, Mexico, or the USSR refrained from cutting the prices of their crudes substantially below $30/b. Among the non-OPEC producers, BNOC was in a precarious position because of the nature of its oil transactions. The terms of participation required BNOC to take participation oil and to sell it back at the market price, either to the North Sea operators or to third-party customers. And because it had no storage facilities, it was forced to persuade the producers to buy back more oil or resort to spot market sales. The activities of BNOC were thus potentially threatening to the survival of the London agreement.

(2) If Libya and Algeria were not tempted to lower their own differentials and thereby forefeit the slight edge given Nigerian crude oil by OPEC.

(3) If Iran, the least predictable and most hawkish of all OPEC members, were to abide by its OPEC-allocated volume of 2.4 million b/d and to avoid undercutting OPEC partners.

To the surprise of all seasoned observers, unlike OPEC's previous experience with production programing, which had lasted only from March to July of 1982, the production programing of 1983 seemed to hold, at least through January 1984. While it is hard to pinpoint all the reasons for the longevity of the latest attempt, several variables seem to play a crucial role. The first variable seemed to be the expectation of recovery in the industrial economies of 1984. In North America, the recovery was expected to gain momentum. In the United States, the cumulative growth in domestic demand was anticipated to be around 10 percent. Even though the prospects of growth in Japan were expected to be somewhat subdued in light of the usual performance of its economy, the expansion in total domestic demand was projected at the annual rate of around 3 percent from mid-1983. In Europe, which continued to be the weakest link, the prospect was for slight recovery between the middle of 1983 and the end of 1984, with output growth to remain below 2 percent in most countries.[63] The second reason for the longevity of the production programing attempt may have been an indication in the second half of 1983 that this strategy to halt market losses was working for OPEC. According to OECD forecasts, the total non-Communist demand for oil was expected to rise from 41.3 million b/d in the first half of 1983 to 44.5 million b/d in the first half of 1984. The strength of the oil market in the second half of 1983 was reflected also in the fact that OPEC's production increased by 16.3 percent over the first half of that year.[64] The third reason appears to have been the slump in spot prices (shown in Table 35), which

Table 35. Selected OPEC and Non-OPEC Spot Crude Oil Prices, October 1982-January 1985

	Saudi Arabia Light API 34°	Libya Zuetina API 40°	Egypt Belayim API 26°	(U.K.) Forties[a] API 36.5°	(Norway) Ekofisk API 42°	(Mexico) Isthmus-Reforma API 34°
1982 Oct.	33.41	34.10	29.00	34.49	34.94	33.19
Nov.	32.03	33.35	28.63	33.15	33.63	32.00
Dec.	30.80	32.30	28.75	31.70	32.20	30.45
1983 Jan.	31.00	31.19	28.13	31.06	31.56	30.31
Feb.	29.56	28.88	27.00	28.84	29.40	28.19
March	28.44	28.00	26.25	28.25	28.75	27.75
April	29.05	29.68	26.63	29.66	30.05	29.06
May	28.65	29.94	26.44	29.42	29.79	28.94
June	28.98	30.05	26.40	n.a.	30.36	29.35
July	29.13	30.46	28.83	n.a.	31.13	29.63
Aug.	28.98	30.71	27.31	n.a.	31.30	29.56
Sept.	28.61	30.27	27.18	30.51[a]	30.64	28.85
Oct.	28.56	29.48	26.80	29.81[a]	29.95	28.65
Nov.	28.28	29.10	26.63	29.19[a]	29.31	n.a.
Dec.	28.26	28.94	26.45	28.69[a]	28.83	n.a.
1984 Jan.	28.64	29.38	26.54	29.53[a]	29.64	n.a.
Feb.	28.61	29.56	26.78	29.89[a]	30.05	n.a.
March	28.57	30.07	27.12	30.06[a]	30.18	n.a.
April	28.45	30.05	27.20	30.13[a]	30.25	n.a.
May	28.43	29.76	27.14	29.81[a]	29.91	n.a.
June	28.12	29.15	27.14	29.35[a]	29.43	n.a.
July	27.72	28.25	26.92	28.72[a]	28.78	n.a.
Aug.	27.79	28.09	26.48	28.50[a]	28.57	n.a.
Sept.	27.94	28.15	26.88	28.53[a]	28.56	n.a.
Oct.	27.85	28.33	26.75	27.88[a]	27.88	n.a.
Nov.	27.96	27.93	26.64	27.80[a]	27.80	n.a.
Dec.	27.78	27.75	26.26	27.05[a]	27.05	n.a.
1985 Jan.	28.08	27.56	26.69	27.05[a]	27.05	n.a.

Source: Various issues of *OPEC Bulletin*, 1982, 1983, 1984, 1985.

[a]Starting in Sept. 1983, prices for U.K. crude are those for Brent 37.4°.

were significant determinants of the pricing behavior of OPEC producers. It should not be inferred from the preceding that all OPEC members stood by the "assurances" they had given at the London meeting of strictly observing production quotas. There were numerous violations, especially by the price hawks. In fact, as a result of quota violations, total OPEC output in October 1983 was reported to be around 18.5 million b/d, 1 million above the ceiling. Toward the end of that year, however, OPEC quota violations declined to 800,000 b/d.

The continued slump in the world oil market toward the end of 1983 once again forced OPEC members to reconsider long-term strategy. In the past, OPEC had considered a pricing formula aimed at introducing real increases in crude prices under buoyant conditions in the industrial economies, and at introducing nominal price reductions under soft market conditions. In 1983, OPEC members appeared to be moving away from such a mechanism. A majority were most concerned about OPEC's shrinking share of the market, which was creating intra-OPEC competitive pressures, forcing the price hawks to violate their quotas, and offer a variety of price discounts to spur sales. One potential means of enhancing their share of the oil market would be a resurgence in demand. Toward this end, the majority of members toyed with freezing the prices in nominal terms for a long duration. The underlying logic was that such a freeze would eventually result in price erosion in real terms, leading to a restoration of demand compatible with the financial needs of OPEC members. Iran, disagreeing with this rationale, advocated annual price increases equivalent to the real rate of interest.[65]

Given the lack of consensus, long-term strategy was tabled for further consideration. Obviously, internal differences were too acute to be resolved in the near future. In fact, in view of growing political dissensions among a number of OPEC states in the 1980s, the prospects of any consensus developing in the near future appeared dim.

Toward the end of 1983, OPEC also saw promise for the coming year if its own members would stay within the alloted production quotas of March 1983, and if there were no more reductions in the prices of the North Sea crudes. Needless to say, the fulfillment of any of these conditions in the coming year was well-nigh impossible.

Forecasts for 1984 and the remainder of the 1980s were optimistic from the perspective of oil consumers and gloomy from the vantage point of OPEC. The U.S. Energy Information Administration (EIA), in its annual energy outlook, predicted "slight declines in the average world price for oil for the next 2-3 years and then . . . a gradual rise in the real price to about the levels of 1980 by the early 1990s."[66] Even though the rate of economic growth for the industrial countries was expected to be 3.6 percent in 1984, the prospects for increased consumption of oil were not very promising. This forecast, according to EIA,

Table 36. Total OECD Oil Intensities, 1973-1982

Year	Industry[a]	Annual Change (%)	Residential/ Commercial[b]	Annual Change (%)	Transportation[c]	Annual Change (%)
1973	100.0	—	100.0	—	100.0	—
1974	95.4	−4.6	89.0	−11.0	99.0	−1.0
1975	93.3	−2.2	84.1	−5.5	102.1	+3.0
1976	99.1	+6.2	88.3	+5.0	100.5	−1.6
1977	94.1	−5.1	86.0	−2.6	102.5	+2.0
1978	91.1	−3.1	79.8	−7.2	104.5	+2.0
1979	90.3	−1.0	76.3	−4.4	101.5	−2.9
1980	79.5	−12.0	72.5	−5.0	96.7	−4.7
1981	70.2	−11.7	65.3	−9.9	92.5	−4.3
1982	65.6	−6.6	60.2	−7.8	92.7	+0.3

Source: *Middle East Economic Survey,* April 16, 1984, p. 16.
[a]Oil Consumption in industry per unit of manufacturing output.
[b]Oil Consumption in the residential/commercial sector per unit of private consumption.
[c]Oil consumption in the transportation sector per unit of GDP (gross domestic production.)

was based partially on the expectation of a warm winter in Europe and the United States. To a large extent, however, the lack of increase in oil consumption in the industrial countries was a result of continued conservation, which was moderating the total demand for energy, particularly petroleum. As shown in Table 36, improvement in energy consumption remained more impressive in industrial sectors than in transportation sectors. Energy consumption in many developing countries was expected to be slow because of their financial difficulties.[67]

Meantime, spot prices remained weak, one of the significant reasons underlying this weakness being the continued violation of the March 1983 production ceiling by OPEC members. In July, when the oil ministers met in Vienna, OPEC production was reported to be 18.7 million b/d, 1.2 million b/d above the ceiling. One purpose of this meeting was to consider Nigeria's request for an increase in its quota from 1.3 to 1.4 million b/d. To accommodate Nigeria, Saudi Arabia agreed to slash its unofficial quota from 5 million to 4.85 million b/d. The second and more significant purpose of the meeting was to persuade OPEC members to abide by the March 1983 production ceilings, and to urge the nonmember producers to avoid production increases. Even though it was obvious by now that most non-OPEC producers were in no mood to cooperate, OPEC refused to stop trying, and for good reason. One of its weakest links, Nigeria, had already established a precedent by unilaterally following the pricing behavior of the North Sea producers; this had led to the first-ever price reduction by OPEC in 1983. By continuing its exhortations to non-OPEC producers, OPEC hoped to keep the March 1983 price reduction from becoming a pattern. What OPEC was only too well aware of, but was equally helpless to prevent, was a variety of cheating techniques continually used by its own members that were creating considerable downward pressure on oil prices.

In the fourth quarter of 1984, "despite rather wistful pronouncements that stability and balance were returning to the market-place," prices in spot and futures markets[68] remained in a state of slump.[69] Whatever surges in demand occurred in the market, the rising production rates from non-OPEC producers, such as Mexico, the Soviet Union, the North Sea producers, and even India and Brazil, fulfilled them. OPEC members' continued trouble in selling their oil was exemplified by the fact that Iraq was able to renew its contract to sell 800,000 b/d to France only after Baghdad agreed to increase the proportion for which prices were governed by spot product netbacks. Similarly, Libyan attempts to sell 200,000 b/d of its crude oil by improving tax terms was only partially successful.[70] Under these circumstances, speculations were increasing that a general rollback in world oil prices had become inevitable. On October 16, the state oil company of Norway, Den Norske Stats Oljeselskap (Statoil) announced a price cut, and Britain followed suit. After reduction, the price of Norwegian crude oil stayed in the range of $28.25-$28.65/b, and the price of British crude oil at $28.40-$28.65/b. Nigeria, in an attempt to protect its share of the oil trade, not only slashed the price of its Bonny Light crude by $2, to $28/b, but, repeating its February 1983 resolve, also expressed its determination to follow North Sea prices down "cent for cent."[71] As an immediate reaction to the price decisions of Norway, Britain, and Nigeria, OPEC was faced with adopting two highly controversial measures. The first of these was a new production ceiling. Even though the 17.5 million b/d ceiling was in force, because of widespread cheating among members the actual production in October was reported to be 18.5 million b/d. In view of recent price cuts, OPEC members were expected to consider a cutback in its ceiling to 16 million b/d. This drastic reduction was expected to trigger acrimonious debates among OPEC members, especially on the touchy issue of individual ceilings. The burden of the production cutback was expected to be absorbed by a reduction of 1.5 million b/d by Saudi Arabia, whose output stood at 4.5 million b/d. Kuwait and Venezuela were also expected to make voluntary cuts. Nigeria, on the other hand, was not expected to be amenable to any suggestions for reduced production.

The second controversial measure faced by OPEC was incorporation of more realistic differentials between light and heavy crudes. Crude oil differentials had been set during the early 1980s, when buyers were willing to pay premium prices for oil, regardless of quality. Even though processing of Arabian Light was not profitable, it was being sold at $29/b, while the price of Arabian Heavy was $26/b.[72]

As expected, OPEC members agreed on a new production ceiling of 16

Table 37. OPEC Production Quotas, October 1984 (1,000 b/d)

Producer	Reduction	New Ceiling
Algeria	62	663
Ecuador	17	183
Gabon	13	137
Indonesia	111	1,189
Iran	100	2,300
Iraq	0	1,200
Kuwait	150	900
Libya	110	990
Nigeria	0	1,300
Qatar	20	280
Saudi Arabia	647	4,353
UAE	150	950
Venezuela	120	1,555
Total	1,500	16,000

Source: *Oil and Gas Journal,* Nov. 5, 1984, p. 58.

million b/d at the conclusion of their Geneva meeting, an overall 8.6 percent cut. This measure was aimed at defending the $29/b price of marker crude (Table 37). With the exception of Nigeria and Iraq, the production volumes of all members were reduced. As usual, Saudi Arabia absorbed the largest cutback as a swing producer. Observers of the oil scene remained skeptical that the OPEC states would abide by these production ceilings.

The sticky question of differentials was handed over to a committee of OPEC ministers, headed by Yamani, for further study. In December it was reported that the Yamani committee wanted to recommend an increase in the prices of heavier crudes while retaining the price of marker crude at $29/b. The oil traders and companies reportedly felt that the market would remain soft until OPEC reduced the price of marker crude. Yamani, on the other hand, expected a tightening of prices once the industry traders, after realizing OPEC was not about to reduce its price, would scramble to build their inventory to meet the late-winter demand.

It was becoming increasingly clear that for a number of reasons, traders and companies were reading the oil market better than the Yamani committee was. First, BNOC and Statoil were orienting their prices toward the market by setting them monthly instead of quarterly. Second, even though two non-OPEC producers, Mexico and Egypt, introduced temporary volume reductions in sympathy with OPEC's efforts to stabilize oil prices,[73] no other non-OPEC producer joined their ranks. Third, given a long-standing record of violating OPEC production quotas, it was virtually certain that a number of OPEC members would continue to undermine OPEC's ceiling by overproducing and underpricing. In fact, it was reported that Iran began offering bigger discounts within a few days of OPEC's October decision to adopt a new production ceiling, and within a few weeks both Nigeria and Iran were reportedly overproducing. As a result OPEC's actual production stayed at 16.5 million b/d. Finally, an examination of spot prices in the fourth quarter of 1984 makes it clear that neither Arabian Light nor Norwegian Ekofisk nor U.K.'s Brent (all major crudes) were being sold above their official prices (Table 35)—a clear indication of a continued soft market.

OPEC members knew that the only way they would be able to stabilize prices was by adhering strictly to the production ceilings of October 1984, in tandem with facing the intricate and contentious problem of differentials. Toward the end of December, OPEC made yet another attempt to discipline its members' production. The production monitoring committee was enlarged to include Abu Dhabi, Algeria, Ecuador, Iran, Iraq, and Libya. A five-member executive council, to be chaired by Yamani, was to police members' production and sales. This council was also authorized to hire outside auditors who would be given complete access to oil-related transactions of the member states. The members further agreed to seek approval of auditing procedures by OPEC heads of state. This move was based on the rationale that such approval would alleviate intragovernmental struggles aimed at increasing production, which was partially responsible for undermining all previous attempts to abide by production ceilings. It was apparent, nonetheless, that even this decision had not gone far enough to stabilize prices. While the monitoring agreement covered production of oil, it did not apply to the control of prices on refined products, which many OPEC members had been selling at reduced prices.

Another problem which OPEC had to come to grips with was the vexing issue of differentials. Heavy crudes had become popular not only because, as a result of upgrading of refineries, they could be refined with good yields, but also because they were outside OPEC's price structure. As a result of the continued decline in the price of Saudi light crude and a similar increase in the price of Saudi heavy oil, the differences between the two prices had narrowed from \$2.80/b in April 1983 to 80¢/b in September 1984.[74] To correct such inequities, the OPEC ministers made an interim adjustment by increasing the price of heavy and medium crudes by 50¢/b and 25¢/b, respectively, and by lowering the price of Arabian Light by 25¢/b. Libya, Nigeria, Qatar, and the UAE argued that the price of heavy crude should be increased by \$1-\$1.50/b. In view of OPEC's refusal to go along with their wishes, Algeria and Nigeria, the two producers of light crude, refused to ratify the interim differentials agreement. The other light oil producers went along with this agreement in the interest of "unity."

This meeting left the marker price of \$29/b intact, at least temporarily. The oil ministers were considering a departure from the use of a specific crude as a marker, toying instead with the idea of using "a basket of crudes with a flexible system of differentials" that could be readily adjusted under changing market conditions. The oil ministers also argued for lowering the marker price, a topic considered sacrosanct only a few months ago.[75]

The interim agreement on differentials was expected to do anything but shore up prices of OPEC crudes. Members kept wary eyes on the actions of the North Sea producers, whose price reductions in October 1984, followed by a similar action by Nigeria, served as a major blow to OPEC's hopes of firming up prices and raising production quotas. While BNOC assured OPEC that it would await the oil market's reaction to OPEC's interim agreement before making any decision on prices for January 1985, BNOC continued to feed as much as 400,000 b/d into spot markets at considerably below the official price of $28.65/b for Brent crude. This action assured confusion in crude oil prices.[76] Norway also was reported to be inching closer to spot pricing.[77]

In the continuing slack market of 1985, in which most OPEC members were finding it increasingly difficult to dispose of their national crude quotas, the issue of differentials remained a source of acrimony and division within the organization. Because the oil states were able to sell their commodity by resorting to deep discounts, adjusting differentials was seen as a way to promote exports and insure that production quotas were not exceeded.[78]

As a producer of light crude, Algeria advocated a $2/b increase in the price of heavy crudes while maintaining the price of marker crude at $29/b. The price of heavy crudes stood at $26.50/b, and incorporation of the Algerian plan would have closed the price difference between light and heavy crudes to about 50¢/b when heavy crudes were the only grades in which official prices were in line with spot levels. The UAE and Qatar, the two Persian Gulf producers of light crude, viewing cheap Nigerian crude as a great threat to the marketability of their own commodity, wanted a $1/b differential. Nigeria, however, wanted OPEC to only consider its own position as a short-haul supplier to Europe in fixing the price. Nigeria also favored a reduction in the price of Arabian Light marker crude to $28.25/b, a readjustment in the heavy price to $27/b, and an increase in the price of its own Murban and Bonny Light crudes to $28.50/b and $28.75/b, respectively. At these levels the prices of Nigerian crudes would be more in line with the official price of one of Nigeria's chief competitors in the European market, Britain, the official price of whose Brent crude was $28.65/b. Heavy- and medium-crude producers were less concerned about competition from Nigeria, and were reportedly more accommodating. Saudi Arabia, on the other hand, was firmly opposed to any upward adjustment in the $26.5/b price of heavy crude.

OPEC members were reported to be edging toward a price reduction in Arabian Light marker crude and were also debating moving away from an OPEC crude as a marker. One view was to use "notional crude" as a marker, but this suggestion was inherently problematic. Since notional crudes were a hypothetical blend and were not traded on spot markets, it would be difficult to determine the market value of such a blend. Another suggestion, which appeared more realistic, was to use a marker crude that was widely available on spot markets and generated a true spot price. Incorporation of this suggestion by OPEC would have assigned the North Sea crudes the status of marker since, as previously noted, Britain and Norway had been allowing spot markets to set the prices of their crudes.

As OPEC continued to ponder these issues, its production fell below 16 million b/d. This may have been due to the willingness of Saudi Arabia to cut back production to less than 3.5 million b/d. Nigeria continued its quota violations, and Iran, as a result of its continuing war, was forced to pay a $2/b discount to compensate for the additional freight and insurance cost of sending tankers to Kharg. In the same period, oil prices in spot and futures markets continued to fall.

Between 1983 and 1985, the actions of the North Sea producers became increasingly significant for OPEC as well as for industry traders. For OPEC, their significance stemmed from the fact that since 1983, Nigerian pricing policy had been determined not by OPEC decisions but by how Britain and Norway were going to respond. Throughout January 1985, it was becoming increasingly clear that, in view of the very large volume of oil coming on spot markets from North Sea producers, the price of oil was destined to go down further. Oil traders seemed to have found a new leverage in Britain and Norway to sustain downward pressure on OPEC prices. OPEC itself, in continuing its long-standing indecision in the prevalence of a soft market, was sending signals that it was getting ready to introduce another price rollback.

On January 8, 1985, by announcing that it would no longer set official prices and would sell its commodity at free market prices, Britain only made its well-established practice official. Even though British officials issued a denial the following day, it was widely known that this purfunctory statement was aimed at appeasing OPEC members, who on numerous occasions had threatened to trigger a price war against Britain and Norway if they continued to create "downward pressures" on OPEC's prices.[79]

As expected, on January 31, OPEC announced its decision to abandon its $29/b benchmark and reduced the price of Arabian Light by $1/b. The prices of medium and heavy crudes were left unchanged at $27.65 and $26.50/b, respectively. Nigeria agreed to raise the price of its Bonny Light crude from $28 to $28.65/b, thus aligning it with the "official" price of U.K.'s Brent crude. Algeria, Libya, and Iran refused to endorse this bitterly fought decision, though Iran later changed its vote. On establishing a monitoring system, the member states demonstrated at least a semblance of unanimity. The organization announced the appointment of an Amsterdam-based international accountant,

KMG Klynveld Kraayenhoff and Company, to monitor production, sales, and pricing of its members.[80] There were no official sanctions for noncompliance, however.

Spot prices hardened a bit in Europe in February with the news that BNOC planned no reduction in its $28.65/b price. North Sea prices reacted to the cold weather in northwestern Europe and also to reduction of crude and distillate shipments from the Soviet Union. This reduction led to unsubstantiated rumors that the Soviet Union had taken this measure deliberately to firm up the market. At the same time, BNOC was continuing to lose money because it had been buying 800,000 b/d of North Sea oil at the official price of $28.65/b but had been forced to sell it at the lower spot prices. Between August 1984 and March 1985, BNOC was reported to have lost $76 million. In January 1985 alone, BNOC lost about $34 million on trading.[81] In March, BNOC was expected to ask the government to subsidize its trading loss of $33 million. In the same month, Britain announced its intention of abolishing BNOC in the spring. The impending demise of BNOC meant that the companies producing North Sea oil would pay taxes and royalties to the government.[82] In its place, Britain was to establish a new organization, the Government Oil and Pipeline agency, which was to perform the functions of BNOC during energy supply emergencies.

The impending demise of BNOC underscored the fact that the spot market had emerged as the "price administrator" of oil. OPEC's objective of retaining control of oil prices, an ongoing effort since 1981, appeared more elusive than ever at the end of the first quarter of 1985. OPEC's price cut also brought to an end, at least temporarily, the cooperation of Egypt and Mexico. The Egyptian oil minister, Abdellah Kandil, who was present at OPEC's last meeting, withdrew because of the strident tone of debates and the resultant impasse. Similarly, Mexico made known its intentions to proceed with independent production and pricing policies. In February, Egypt and Mexico reduced prices of their crudes to $25.50 and $25.75/b, respectively, and in March, Britain made an unequivocal decision to stop setting the price of its North Sea oil. If the Egyptian and Mexican decisions were expressions of their lack of faith in the capability of OPEC members to abide by the October 1984 production ceiling agreement, then the British move indicated its pessimism at the possibility of long-term strength in oil prices in the foreseeable future.

Intra-OPEC Tensions

OPEC was never crippled by the political differences among its members. The three dominant members—Saudi Arabia, Iran, and Iraq—managed to find common points of agreement on crude oil prices throughout the 1960s and 1970s, though this consensus-building was not an easy task. In the 1980s, however, political and military tensions among OPEC members appeared to be growing at a faster pace. If unchecked, these tensions might lead not only to potential dismantlement of OPEC, but, worse than that, to a major regional or even a superpower conflict. Two of these tensions were the struggle over the leadership of the organization and the Khomeini factor.

The leadership of OPEC has never been a simple phenomenon to be determined solely on the basis of which country owned the largest oil reserves. The ability of an OPEC member to dominate the region both politically and militarily has also been a crucial variable. And since no one country has been a dominant force in all three areas, the issue of leadership of OPEC in the 1970s became a struggle between two major actors: Saudi Arabia, which owned the largest oil reserves, and Iran, which, in addition to being one of the largest producers of oil, remained during that decade the dominant political and military force in the Persian Gulf. The modality of this power contest between Iran and Saudi Arabia was clearly established in the 1970s when these two countries spearheaded the hawkish and moderate groups, respectively within OPEC. The issue of contention between the two groups was the timing and amount of price escalations. As the largest oil producer, and as a country concerned with the ramifications of intermittent quantum price escalations for the sustained growth of the interdependent non-Communist economic system, Saudi Arabia remained a persistent advocate of a price freeze between 1974 and 1976 and of moderate increases between 1976 and 1978. Conversely, Iran, by pressing for larger increases, gained wider support within OPEC. The dynamics of intra-OPEC struggle transcended the conventional differences between monarchical (conservative) and revolutionary (radical) regimes. The price hawks included Iran (a monarchy until 1978), Libya (a firebrand revolutionary since 1969), Algeria (revolutionary), and Nigeria (a democracy frequently interrupted by military intervention), which were periodically joined by Iraq (a Bathist revolutionary), Indonesia (a right-wing military dictatorship), and Venezuela (a democracy). It should be noted that in the struggle for leadership of OPEC, Saudi Arabia, and Iran should have been more conscious of the implications of this struggle for the future dynamics of OPEC decisions. It is unclear whether, in aligning their support around Iran or Saudi Arabia, other OPEC members were equally conscious of this particular aspect of the struggle. Observing the behavior of OPEC throughout the 1970s, one can surmise that most OPEC members were so preoccupied with the ultimate prevalence of their respective pricing positions at a given time that the long-term implications of this struggle became almost immaterial to them. It should also be emphasized that the struggle over the leadership of OPEC remained a dormant issue under the tight market of 1979-1980. It was only in soft market periods (for instance, 1974-1978 and 1981-1985) that this struggle created considerable intra-OPEC tensions. The soft market also enhanced the capability of Saudi Arabia to impose its will on the price hawks because of the sheer weight of its reserves. The December 1976 conference, which resulted in OPEC's first two-tier price system, was a case in which the Saudi wish to force the majority of OPEC to lower prices did not prevail because the oil market remained buoyant. In 1981, however, Saudi Arabia, under a prolonged soft market, for the first time succeeded in imposing its will on the other OPEC members, forcing them to realign prices at the lowered Saudi scale.

The struggle between Iran and Saudi Arabia over the leadership of OPEC continued in the 1980s. In 1982, as OPEC was gearing up to tackle production programing, Iran argued that allocation of quotas be based on population and financial need. Such a formula clearly put Iran (population 30 million), Nigeria (80 million), Indonesia (150 million), and even Iraq (13 million) ahead of Saudi Arabia (5 million), since all populous states, as "high absorbers" of capital, were also in acute need of capital. Needless to say, the populous oil states were quite sympathetic to the Iranian argument, especially when they considered that Saudi Arabia was one of the leading "low absorbers" (see Table A-8). Acceptance of the Iranian position by the Saudis, nevertheless, would clearly enable Iran to regain the power within OPEC it had enjoyed under the Shah, though it would not guarantee that Iran would stop its struggle for dominance within OPEC in the 1980s. Moreover, the political implications of enhanced leadership status for Iran might further escalate the already threatening posture of the Khomeini regime in the Persian Gulf, a repulsive scenario not only for the Saudis but also for other oil monarchies.

When Iran was accused of violating its March 1982 quota of 1.2 million b/d and undercutting other OPEC members by selling its crude oil at a lower price, the Iranian oil minister, Mohammed Gharazi, reminded his OPEC colleagues that they should, in fact, be grateful to Iran for drastically lowering its 6 million b/d production during the days of the Shah to about 2 million b/d. Iran accused Saudi Arabia of unfairly expanding its share at the expense of Iran at a time when the latter was fighting a war with Iraq. Under such circumstances, Iran was left with no alternative but to exploit whatever opportunities presented themselves. Consequently, Iran remained the leading violator of the OPEC-allocated quotas of 1982. When the March 1983 production ceiling agreement was negotiated, Gharazi publicly endorsed the move by accepting the OPEC-assigned quota of 2.4 million b/d. There seemed to be three significant reasons underlying the Iranian cooperative behavior. First, it was quite apparent that a probable alternative to cartelization was the collapse of OPEC, a most undesirable scenario for all OPEC states. The second reason appeared to be that the volume assigned to Iran in 1982 (2.4 million b/d) was closer to that country's economic needs than the 1.2 million b/d assigned in 1982. The third reason for Iranian cooperation seemed to be OPEC's assigning of Saudi Arabia the status of a swing producer in 1983 (rather than any specific quota), thereby diffusing a

Table 38. Estimated OPEC Oil Revenues, 1974-1983 ($ billion)

	1974	1979	1980	1981	1982	1983[a]	% of Change 1982-1983
Saudi Arabia	22.6	57.5	102.0	113.2	76.0	46.1	−39
UAE	5.5	12.9	19.5	18.7	16.0	12.8	−20
Kuwait	7.0	16.7	17.9	14.9	10.0	9.9	−1
Iran	17.5	19.1	13.5	8.6	19.0	21.7	+14
Iraq	5.7	21.3	26.0	10.4	9.5	8.4	−12
Qatar	1.6	3.6	5.4	5.3	4.2	3.0	−29
Nigeria	8.9	16.6	25.6	18.3	14.0	10.0	−28
Libya	6.0	15.2	22.6	15.6	14.0	11.2	−20
Algeria	3.7	7.5	12.5	10.8	8.5	9.7	+14
Venezuela	8.7	13.5	17.6	19.9	16.5	15.0	−9
Indonesia	3.3	8.9	12.9	14.1	11.5	9.9	−14
Gabon	n.a.	1.4	1.8	1.6	1.5	1.5	—
Ecuador	n.a.	1.0	1.4	1.5	1.2	1.1	−8
Total OPEC	90.5	195.2	278.8	252.9	201.9	160.4	−21

Source: *Petroleum Economist,* June 1984.
[a]Provisional

main source of contention for Iran. Iran also accepted the production quota of October 1984. Using the contingencies of war as a rationale, however, the Iranian government continued to violate its production ceiling and offer discounts to its customers.

In the early 1980s, Saudi Arabia clearly emerged as the leader of OPEC. In this capacity, it continues to bear the brunt of production cutbacks, not only to minimize the effects of downward pressures on the price of OPEC oil, but also to help such countries as Nigeria whose revenues have been substantially declining since 1980 (see Table 38) as a result of the continued slump in the world oil market and increased price competition from the U.K. and Norway. Table 38 also shows a progressive decline in the Saudi oil revenues since 1981: from $113.2 billion in 1981 to $76 billion in 1982 (a decrease of 33 percent) and to $46.1 billion in 1983 (a decrease of 39 percent from 1982 and 59 percent from 1981). It should be noted, however, that the emergence of Saudi Arabia as the leader of OPEC is only a temporary phenomenon because the continuation of the Iran-Iraq war has kept this issue more or less on a back burner. The outcome of this war is most likely to provide a new impetus for this struggle. If a clear winner emerges, then that actor, perceiving itself as a natural candidate for the role of dominant actor (à la Iran under the Shah), is also likely to reinvigorate the contest for the leadership of OPEC. If the victor is Iraq, the leadership struggle is likely to be more contained, similar to the one between Saudi Arabia and Iran under the Shah in the 1970s. If the victor is Khomeini's Iran, the struggle is likely to be fought via "guerilla warfare," a kind of hit-and-run contest between the Saudis and the Iranians in political, economic, and even military affairs, a battle which might even threaten the prolonged survival of OPEC.

Since its accession to power, the Khomeini regime has been a constant source of consternation to the conservative oil monarchies and a serious threat to regional political stability. Iran has been at war with Iraq since 1980. Even though Iran and Iraq are the only two direct parties to this conflict, other countries of the region have become indirectly involved in order to maximize their respective strategic gains from the war's outcome. For instance, Qaddafi in Libya, who has long envisioned himself as heir to the legacy of the late Gamal Abdel Nasser of Egypt and his notions of Pan Arabism, is siding with Iran. Because Egypt is still treated more or less as a pariah in the Arab world because of its peace treaty with Israel, Qaddafi perceives Iraq and Syria as two sources of threat to his ambitions for strategic dominance in the region. Syria is entangled too deeply in Lebanon to become a dominant force. And, as Qaddafi sees it, the present political realities in Lebanon indicate that even after the Israeli withdrawal from southern Lebanon, the Syrian responsibilities in Lebanon are going to be too intricate for President Hafez Assad to fulfill his own ambitions of emerging as one of the dominant forces in the Middle East. The process of elimination leaves Iraq as the only potential Arab challenger to Qaddafi. The Iran/Iraq war does not appear to undermine Qaddafi's strategic objectives. The longer and harder Iran and Iraq inflict damage on one another, the longer and more difficult it will be for either of them to become a dominant regional actor, even after settlement of the war. A victory for Iran may not be a bad scenario for Qaddafi, since a victorious Khomeini would be more of a threat to other power contenders such as the Saudis. A decisive victory for Iraq remains an unthinkable scenario for Libya. Syria also sides with Iran. Syrian motives, like those of Libya, are based upon President Assad's desire to see a much weakened posture for Iraq, which would enable Assad to emerge as a dominant force within the Bathist movement, the rival wings of which now rule Iraq and Syria.

Saudi Arabia and other Persian Gulf sheikhdoms not only side with Iraq but also have invested billions of dollars in the Iraqi war chest. Such a heavy commitment can be understood if one examines the political posture of Iran under Khomeini. The Khomeini regime has been consistently driven by a religious zeal and superciliousness that is unprecedented in modern times. The religious fervor of this regime makes it eager not only to pass self-righteous judgment on all entities within its environment but also to strive to transform the present political status in its own image. Thus, neighboring governments are not considered "true adherents" of Islam in the Khomeini tradition, substantially because of their friendly posture toward the United States, which the Khomeini regime brands the "Great Satan." According to this logic, the friends of Satan cannot claim to be faithful to Islam and must be replaced by followers of the Khomeini orthodoxy.[83] As a consequence, the Persian Gulf monarchies appear to be suffering from a paranoia reminiscent of the 1950s and 1960s, when Nasserite elements allegedly attempted to overthrow the monarchical regimes. The defeat of Iran in this war is most likely to diminish the threatening posture of Khomeini's Iran vis-à-vis the monarchies. In addition, the weakening of the Khomeini regime might not only discourage a variety of power contenders within the Persian Gulf from attempting to stage a Khomeini-style revolution elsewhere but would also be a first step, the oil monarchies hope, toward its eventual uprooting from Iran.

The intransigence of the Khomeini regime concerning a negotiated settlement seems to have thwarted all attempts by the oil monarchies and Iraq to end this war. Iran, under Khomeini, rejects the notion of compromise in any of its manifestations and sees compromise as an "unholy" word. Thus the Khomeini regime has insisted all along that it will not negotiate a settlement of the war unless Iraq accepts the blame for starting it, relinquishes all captured territory, and pays reparations for war damage. At times Iran has denied any need for negotiation with Iraq, proclaiming Iran's "greatest right" is to overthrow the regime of Saddam Hussein in Iraq, and has maintained that "any final settlement must be done in the Islamic context to judge and punish the aggressor." In the face of this inflexibility by Iran, the Saudis have two policy options: to continue all efforts to seek a peaceful resolution of the war, and, as assurance against the potential dismantlement of the Saddam regime, to continue to provide economic assistance and political support for Iraq. Needless to say, Saudi Arabia is pursuing both options with equal vigor.

One additional aspect of the Saudi role vis-à-vis the Iran–Iraq war must also be considered in the light of equally substantial financial support for Syria. The fact that Syria and Iraq are at loggerheads seems to support the hope of Saudi Arabia to become a dominant regional force. If the Iran-Iraq war is settled decisively in favor of Iraq, Saudi Arabia will have to be concerned with a rejuvenation of antagonistic relations with Iraq, whose irredentist claims on Kuwait and support for the Marxist guerrillas in Oman's Dhofar Province in the 1960s and early 1970s are only too fresh in the memory of the Saudi monarchy.[84] This is where the Saudis can count on utilizing the "Syrian card" against Iraq to insulate themselves from potentially aggressive Iraqi designs. Thus, regarding the Iran-Iraq war, the Saudi monarchy appears to be exercising the prudence for which it has been well-known among observers. This circumspection notwithstanding, Saudi Arabia appears bewildered by the paradoxical posturing of Iran within and outside of OPEC.

If one examines the behavior of Iran under Khomeini, it seems there are two Irans: an "economic Iran," operating within the framework of OPEC, and a "political Iran," operating outside OPEC. The economic Iran appears pragmatic and willing to cooperate with other oil states on matters of economic interest, including Iraq. In fact, as previously noted, despite its strident political posture, Iran cooperated with the Gulf states on incorporating production programing in 1982, 1983, and 1984. At the same time, however, in light of pressing economic needs stemming primarily from the ongoing war, Iran continued to violate production programing during all three years.

The political Iran, on the contrary, appears uncompromising, unyielding, strident, ideological, and war-mongering. The chief threat to the cohesion and even the longevity of OPEC stems from the threatening posture of the political Iran. The continuing war between Iran and Iraq and the partisan association of the Persian Gulf states vis-à-vis this war are two of the least predictable and most explosive variables that have the potential of converting an indecisive series of battles into a regional war. And the chances are that, once triggered, such a war might not be limited to military actors with limited firepower. Both superpowers are also in the region with their ominous nuclear might lingering in the background. A possible expansion of the war between Iran and Iraq contains all the ingredients of a calamity for the human race.

From 1976 through the first quarter of 1985 OPEC continued to ride the roller coaster of oil power. Even though these countries enjoyed the status of advantaged actors between 1976 and 1980, their ability to manipulate the oil market in the 1980s has definitely dissipated.

In general, the oil market looked promising for OPEC between 1976 and 1978. It was divergent interpretations of market conditions by the moderates and the hawks that led to the first two-tier price system. The Saudis, determined to give the OECD economies ample time to recuperate from the shocks of 1973, wanted minimal price increases, and the price hawks felt that the economic recovery was strong enough to absorb a comparatively higher increase without fear of setback. The continued buoyancy of the market, despite Saudi Arabia's endeavors to flex its muscle as a swing producer, worked in favor of the price hawks.

The Iranian revolution provided perhaps the last opportunity for the OPEC members to send shockwaves through the Western economies by introducing a string of unrealistic price escalations. From the schedules of price increases agreed upon in December 1978 until the conclusion of 1980, OPEC members established a new record of price leapfrogging, with some price increases only weeks or hours apart. The following is an overview of this period:

(1) In a December 1978 meeting, OPEC members agreed upon price schedules for the coming year whereby the price of Arabian Light marker crude was to be raised from $12.700 to $14.542/b in four installments by the last quarter of 1979.[85]

(2) In March 1979, OPEC decided to apply the $14.542/b scale that was originally scheduled for the fourth quarter of that year.

(3) In June, OPEC agreed on a new pricing arrangement whereby the price of Arabian Light was set at $18/b, Iranian Light was established at $21-22/b, and a ceiling of $23.50/b was agreed upon for African crude.

(4) In July, Qatar breached the spirit of the preceding arrangement by selling crude at $34-45/b. In the next three months, Nigeria, Kuwait, Iran, Libya, Iraq, and Algeria raised their prices beyond the June ceiling of $23.50/b.

(5) At the conclusion of OPEC's December meeting, Saudi Arabia raised the price of Arabian Light crude to $24/b in order to align it with the price charged by Venezuela, Indonesia, Iraq, Kuwait, the UAE, and Qatar. Algeria, Libya, and Nigeria raised their prices to $30/b.

(6) Within one week of the preceding conference, Venezuela, Indonesia, Iraq, Kuwait, the UAE, and Qatar raised their prices to $26/b.

(7) On January 28, 1980, Saudi Arabia realigned the price of Arabian Light at $26/b, effective January 1.

(8) Within twenty-four hours, the countries listed under #6 raised their prices to $28/b.

(9) In May, Saudi Arabia once again realigned its price at $28/b.

(10) In less than a week, the countries listed under #6 once again escalated their price to $30/b.

(11) In August, Saudi Arabia was persuaded to realign the price of Arabian Light at $30/b.

(12) At the December OPEC conference, following the outbreak of hostilities between Iran and Iraq, Saudi Arabia raised its price to $32/b.

(13) In January 1981, the countries listed under #6 raised the price of crudes from $32/b to $36/b.

(14) In October 1981, OPEC lowered its price and reunited it at $34/b.

One of the most significant and long-lasting outcomes of these price escalations was that OPEC unwittingly abdicated its role as a price administrator, starting a new trend of following spot prices. An immediate consequence was the disintegration of a unified price structure around the marker crude. Instead, a four-tier price structure prevailed from time to time in 1979 and 1980.

Special mention must be made of the Saudi Arabian role in the 1980s. The emergence of a soft market in 1981 guaranteed reinstatement of that country's role as a swing producer. This time, however, unlike 1977, the slump in the market was pervasive enough to allow the Saudi-engineered oil glut, coupled with lower prices of Saudi crude, to force the other OPEC members to lower their prices in order to reunify them at $34/b, a scale desired by the Saudis. But when OPEC's price structure came under intense downward pressure from the continued slump in the OECD economies, Saudi Arabia became one of the leading advocates of preventive measures. The March 1982 production programing arrangement, which enabled OPEC to operate as a true cartel at least for a short time, was the outcome of Saudi endeavors. After the collapse of this arrangement, Saudi Arabia, along with the Gulf states, continued its quest to keep the OPEC price structure from collapsing under the combined effects of growing competition from non-OPEC producers, a continued recession, and institutionalized conservation measures and fuel-switching in the West. The two-pronged OPEC strategy developed in March 1983, which included production programing along with a first-ever reduction in the official price of OPEC marker crude, was the outcome of this quest. The significance attached by Saudi Arabia to the March 1983 arrangement became apparent only in the fact that that country, as a swing producer, was willing to bear the brunt of ensuing production cutbacks.

Non-OPEC producers emerged as one of the foremost sources of uncertainty for OPEC's ability to operate as a cartel. At the March 1983 conference, for the first time in its history OPEC was forced not only to seek cooperation from non-OPEC producers on the issue of pricing but also to ponder the need for sustained cooperation with them. Throughout 1983, OPEC's production programing was not threatened by the volume-maximizing and price-undercutting activities of non-OPEC producers. But the possibility remained that non-OPEC producers might take measures, however unwittingly, that would cause disarray within the rank and file of OPEC. The October 1984 production programing arrangement as well as the January 1985 price reduction decision were results of continued volume maximization in tandem with spot market trading by the North Sea producers.

The long-term strategy that OPEC intermittently considered in the 1970s was not bold enough to maintain its status as a price administrator. The transformation of OPEC into a residual producer in the 1980s made this strategy archaic and irrelevant. In order to survive in the 1980s, OPEC must continue to slash its production rates and roll back prices. Considering the fact that about 90 percent of the governmental revenues of almost all OPEC members come from oil, the petroleum exporting states cannot be expected to sustain drastic production cutbacks or continue to roll back prices of their commodity. The continuance of a soft market might be disastrous for the longevity of that organization.

The issue of intra-OPEC political dissensions is disturbing for the political stability of the Persian Gulf. Thus far, Iran has maintained an overall cooperative posture on economic matters, but the political differences between the Khomeini regime and the conservative Arab monarchies, and the continuing hostilities between Iran and Iraq are potentially explosive and may yet undermine the unity and cohesion of OPEC.

7. Epilogue

To sum up the dynamics of oil affairs between the petroleum-exporting and the industrialized consuming countries, the main theme of this study must be restated: that, although political considerations may at times have entered into the calculations of the oil states, OPEC's pricing behavior has been based essentially on economic considerations. This study has also underscored the fact that bargaining between oil states and oil companies, between oil-producing and oil-consuming nations, and even among oil states themselves has remained acutely subservient to the uncertainties of the marketplace. When the oil states were victimized by the buyers's market in the 1960s, they aspired to raise their revenues by negotiating minimal increases in crude oil prices. Their attempts, however, met with little success. The oil companies (and, by extension, the industrial consuming nations) were in such an advantaged position that they saw no compelling reason to bargain with their tremendously disadvantaged opponents. The unwillingness of the oil companies to cooperate was perceived by the petroleum-exporting states as an unrealistic negotiating stance, and no real basis for cooperation was ever established between them.[1] This lack of cooperation in itself assured sustenance of a similar relationship between the two groups when the oil states became the advantaged actors and gleefully dismantled the gains made by the oil industry, which they regarded as illegitimate.

The ruthless and aggressive bargaining posture of the oil states vis-à-vis the oil companies became increasingly pronounced as the pendulum of advantage in a growing sellers' market appeared to swing away from the oil companies between 1971 and 1973. In this period, the nature of the oil states' demands underwent a process of radicalization. The Tripoli I, Teheran, and Tripoli II negotiations were concerned with straightforward increases in crude oil prices, the inflation-related demands being most apparent in the Teheran agreement. In the Geneva I and II agreements, the concern of OPEC members broadened in the hope of recouping the loss of their buying power stemming from recent dollar devaluations. Meantime, their demands for phased participation in upstream operations—the most radical demands until then—were also agreed upon in October 1972. Thanks to tremendous strides in the oil states' bargaining position in the ensuing period, however, these agreements were never implemented in their original form. The gleeful attitude of the oil states in setting aside these agreements and concomitant demands for immediate 51 percent participation created an environment in which even the Saudis could not implement the original agreement without providing fuel for their critics' claims that they were appeasing the Western oil companies and their parent countries. It should be noted that, despite the predominantly economic character of the participation negotiations, the very demand that the oil states control their upstream operations was a reflection of their growing political consciousness. The participation agreements themselves were not only the clearest evidence of OPEC's increasing power but also perhaps indications of further radicalization of their demands in the near future. Such a radicalization became apparent toward the end of 1973 in OPEC's refusal to negotiate price increases with the oil industry.

When the oil states did away with the long-standing tradition of negotiated price increases in the fourth quarter of 1973, it was expected that a basis for cooperation might be found between the governments of oil-producing and oil-consuming nations, since the congruity of objectives of these governments could be delineated much more easily than those of the oil states and the oil companies. There is no doubt that the two groups of governments recognized the necessity for cooperation. But such a recognition has yet to produce a sustained dialogue aimed at finding solutions to some of the most intractable economic problems of our time. The burden of this failure must be borne by OPEC and its Third World allies. Their preoccupation with the persistent and inordinate "unfairness" of their asymmetric economic relations with the West prevented them from developing realistic demands. This inability of OPEC and its allies seems to have been substantially related to their exaggerated perception of the ability of oil power to coax the industrial consumers into yielding to demands for creation of the so-called New International Economic Order. The Western industrial consuming nations saw it as patently unfair for OPEC and its allies to use a vital commodity such as oil to coerce the West into conceding on the NIEO-related demands. Both sides were so convinced of the simultaneous correctness of their respective positions and of the unjustness of their opponents' negotiating stance that the idea of give-and-take, which is an integral aspect of any collective bargaining situation, was drowned in the outpouring of emotional rhetoric from both sides. Thus, Western leaders and political observers made (and continue to make) frequent references to the "moral consequences of giving in to OPECs blackmail," to "the oil shakedown," and so forth. The leaders of the oil states and their Third World allies, on the other hand, frequently referred (and continue to refer) to "the Western exploiters," "neo-colonialism," and "neo-imperialism." Although in the short run such phraseology may have served as effective "condensation symbols,"[2] useful in promoting the emotive aspects of in-group cohesion on both sides, the acrimony stemming from their use may also have been one of the chief obstacles to development of realistic and rational bases of cooperation between OPEC and its allies and the West. In the meantime, each group maintained a confrontational posture and continued to wait for an opportune moment that would so drastically improve its own position and so deteriorate the position of the other side that it could decisively gain lost ground. Under such circumstances, which side is the advantaged actor at a given time is immaterial, since the gains made by the advantaged actor are bound to be unraveled by the disadvantaged actor when the roles are reversed.

The political and economic aspects of oil power did not provide the oil states with boundless ability to influence events or control their own destiny. If anything, the limitations and constraints of the political and economic aspects of oil power must have been glaringly obvious throughout this study. The Arab oil embargo of 1973-1974, in addition to being an excellent example of the translation of economic influence into political clout by the Arab oil states, is a clear example of the limitations of the political aspects of oil power. It is worth repeating that the oil embargo became a reality only because it was a highly profitable venture for its implementors. The profitability of the embargo was underscored by the fact that no participating Arab oil state suffered any loss of revenue. Even the decision of Iraq to participate partially was a good indicator that economic, not political, variables played an overriding role in determining which oil states participated and which did not. Total participation would have proved uneconomical, while a total lack of participation would have been politically suicidal for the Bathist regime in the charged political environment then prevailing in the Arab Middle East. Iraq decided to capture the best results of participation as well as of nonparticipation. By partially participating in the oil embargo, the Iraqi regime adopted a political course guaranteeing its own survival while continuing to reap a bonanza by not introducing drastic cutbacks in its rate of production.

By imposing the oil embargo, the Arab oil states succeeded in bringing about noticeable changes in the policies of the Common Market countries and Japan concerning the Arab/Israeli conflict. But the limitations of the political aspects of oil power were evident in the fact that OAPEC members, aside from unprecedented dramatizing and publicizing of the need for resolution of the Arab/Israeli conflict, did not succeed in bringing about Israeli withdrawal from the occupied territories.

The economic ability to coax the oil companies into signing the Tripoli I and II, Teheran, and Geneva I and II agreements remained totally subservient to the vagaries of the international oil market. The limitations of economic influence became particularly obvious when the oil states continued to exercise it routinely in the 1970s. As advantaged actors, when it was no longer problematic for them to escalate crude prices, the oil states encountered a new set of uncertainties: how to minimize the erosive impact of international inflation and currency fluctuation on their buying power. By acquiring riches, they also enhanced their vulnerability to the booms and busts of non-Communist economies. Since their economies were not sophisticated enough to introduce corrective measures (slowing down economic activity under high inflationary conditions or boosting it under recessionary conditions), the oil states could use the only meaningful weapon at their disposal, tinkering with the prices of their commodity. Consequently, oil prices went through the following discernible phases in the 1970s and early 1980s:

(1) Tight market conditions were worsened by supply shortages stemming either from a deliberate political action (the oil embargo of 1973-1974) or from a political crisis (the Iranian revolution in 1978).

(2) The oil states responded by raising production (Iran, Nigeria, and Indonesia during the oil embargo; Saudi Arabia and other oil states after the Iranian revolution) and by initiating a series of price escalations (immediately preceding and during the oil embargo in 1973 and in 1979 and 1980).

(3) In a tight market, Saudi Arabia, joined by minor actors such as the UAE and other oil sheikhdoms, attempted to moderate price increases, generally without success (1977, 1979, 1980).

(4) Faced with price increases, the industrial consumers adopted two courses of action. First, they came up with additional capital to pay their oil bills, the short-run effect of which was increased inflation (1971-1973) until a saturation point was reached. Then, unable to pay continued price increases, they introduced fiscal and monetary policies in the medium range, thereby slowing down their economic activity (1974-1976 and 1981-1984). They also adopted conservation-oriented policies, the medium- and long-range effects of which increased the energy efficiency of their industrial, transportation, and residential sectors (1981-1985).

(5) The economic downturn reestablished Saudi Arabia as a force disciplining the behavior of the OPEC price hawks and created downward pressure on the OPEC price structure. The price hawks demanded reduced production aimed at ameliorating downward pressure, but met with one of three results: no response (crude oil prices were frozen, 1974-1976); moderate increases (1976-1978); or a two-tier price mechanism with a slight increase in the buoyancy of the market (1977).

(6) The combination of a sustained economic slump and a soft market enabled Saudi Arabia to use its power as a swing producer to force the majority of OPEC members to lower prices (1981). But before other OPEC members lowered their prices, the following phenomena also occurred: (a) Export vol-

umes were reduced for the relatively overpriced crudes from African producers—Libya, Algeria, and Nigeria. (b) The African producers first scaled down their differentials, then eliminated them entirely. (c) Under substantial surplus conditions, these producers were left with no choice but to meet the Saudi demands for price unifications by lowering their own.

The Iranian revolution seems to have been the last opportunity for OPEC to initiate a series of unrealistic price increases. When a soft market finally emerged toward the end of 1981, OPEC was no longer the administrator of prices but was forced to follow the lead of spot prices, establishing a strong tradition. In the 1980s, other realities also worked to assign primacy to the spot market. One such factor was the greatly diminished role of the international oil companies. In 1973 the seven international majors controlled 90 percent of the world oil trade, and their control remained at around 75 percent prior to the Iranian revolution. By 1980, however, their share had dwindled to below 40 percent. The oligopolistic control of the oil market by the international majors of the early 1970s was replaced by a hyperpluralistic oil market in which a large number of purchasers were competing for the best possible deals. This hyperpluralism must be viewed also in conjunction with the huge inventory building by consumers and the low popularity of long-term contracts after the Iranian revolution.

A bitter lesson learned by consumers and by other oil entrepreneurs after the political revolution in Iran was to rely heavily on their own private stocks of oil as a buffer against disruptions of supply. Oil inventories were built up to unprecedentedly high levels during and after 1979-1980.[3] So crucial was the impact of this buildup that it was largely responsible for triggering a price explosion in that period; by the same token, the stock drawdown during 1981-1982 aggravated existing slack market conditions, creating intense downward pressure on OPEC prices.

The nature of oil contracts also went through dramatic mutations during the Iranian revolution, making prices more volatile. In most instances, six-month and one-year contracts were shortened by producers to ninety days or even less, and credit terms were often reduced from ninety to thirty days. The oil thus released was brought on the spot market. The tradition, once established by the oil-exporting countries, was routinized by the oil traders (including many major companies) in the soft market of 1981-1985, when they manifested a distinct preference for short-term contracts. Now the oil states were striving for long-term contracts rather than short-term transactions. In the 1980s, the dwindling role of long-term contracts, and the downward slide of crude oil prices also has enabled the futures market to emerge as a significant trading place. While the futures market served as a cushion for oil traders and speculators, it did nothing to strengthen oil prices, which remained one of the chief preoccupations of OPEC members between 1981 and 1985.

Table 39. Dynamics of the World Crude Oil Market, 1970s-1980s

Market Factor	Characteristics Impact in the 1970s	Current or Expected Characteristics/Impact
Demand for oil	Rapid growth	Flat or very slow growth
Refinery capacity	High utilization rates, very little upgrading; growing demand for light sweet crudes	Huge surplus, many upgradings, declining demand for light sweet crudes.
Quality differentials	Rising; price issue was secondary to supply availability	Falling/low; supply and availability issues secondary to price.
Spot or noncontract markets	Low volume; spot prices exceeded contract prices; OPEC followed spot to raise its term prices; spot market was an expensive source of supply; oil companies wanted supply assurance through term contracts	Very high volume and OPEC is very active in this market; spot prices below contract and reflect the true value of the quality differentials; OPEC reluctant to follow spot to lower term prices; spot market as a cheaper source of supply; oil companies have to meet competition by cutting their crude cost; supply availability not an issue
Revenue needs	Most OPEC countries had current account surplus	Most OPEC countries have current account deficit
Excess production capacity	Little or no excess production capacity; most OPEC countries wanted to produce less due to problems they had with the absorption of oil revenues	Large amounts of excess production capacity; all members are eager to produce more
Futures market for oil	Did not exist	Provides ample information which flows directly to the spot markets
OPEC export refineries	Limited and mostly controlled by the international major oil companies	Growing in importance; generally considered as destabilizing, particularly if product prices are subsidized by lower transfer prices for the feedstock

Table 39. (continued)

Market Factor	Characteristics Impact in the 1970s	Current or Expected Characteristics/Impact
Short term expectations	Prices were expected to go up; continuously anticipated shortages or supply interruptions; inventories were built at all levels for operational and speculative purposes	Prices will come down; no shortages anticipated; interruptions could only cause a temporary and mild increase in the prices; stock drawdowns and inventory optimizations are widely used

Source: *Oil and Gas Journal*, Dec. 31, 1984, p. 46.

The role of non-OPEC producers also caused considerable consternation within OPEC membership. Even in the slack market conditions of 1981-1985, these producers remained almost oblivious to the effects of their volume-maximization on the non-Communist price structure. Even though OPEC's share of the oil market dwindled from about 60 percent in 1978 to 32.1 percent in 1984,[4] that organization continued to bear the brunt of price stabilization.

The cumulative effect of hyperpluralism in the oil market, purchasers' sustained manipulation of stocking/destocking and the related pressures on oil prices, the unpopularity of long-term oil contracts, and the destabilizing role (from OPEC's perspective) of non-OPEC producers seemed to give primacy to the spot and futures markets as the determiners of oil price. An overview of how the crude oil market has changed from the 1970s to 1980s is provided in Table 39.

One can conclude from the preceding that OPEC must come to terms with its new role of the 1980s as the follower of spot prices and, in view of continued volume-maximization by non-OPEC producers, as a residual producer. This task appears to be an arduous one. The members of OPEC enjoyed considerable maneuverability, at least between 1971 and 1973, and again between 1979 and 1980. The period from 1974 through 1978 caused little hardship for OPEC (even though the average oil price of its member states declined by about 11 percent in real terms), especially when compared to the years between 1982 and the end of the first quarter of 1985. The realities of the period since early 1981 indicate that, at least for the foreseeable future, the greatest task faced by OPEC is to attempt to create a sustained calm in the spot market. Thus far, OPEC has attempted to incorporate production programing to stabilize oil prices twice (March 1982 and March 1983) and price reduction once (March 1983) without success. Even though the third production programing attempt and the second

Table 40. OPEC Oil Capacity and Production, 1984

Producer	Production Capacity (million b/d)	1984 Production (1,000 b/d)	Percent of Capacity
Algeria	1.0	608	60.8
Indonesia	1.5	1,423	70.2
Iran	4.0	2,174	54.3
Iraq	4.0	1,210	30.2
Kuwait	3.0	937	31.2
Libya	2.0	1,091	54.5
Nigeria	2.0	1,414	70.5
Saudi Arabia	12.0	4,464	37.2
UAE	2.0	1,159	57.9
Venezuela	2.5	1,711	68.4
Ecuador[a]		258	
Gabon[a]		150	
Qatar[a]		396	
Neutral Zone	n.a.	439	
Total	35.0	17,434	49.8

Sources: 1984 production figures are from *Oil and Gas Journal*, March 11, 1985; production capacity figures are from *Middle East Economic Survey*, Sept. 29, 1980.

[a]Since the combined production capacities of Gabon, Ecuador, and Qatar total 1 million b/d, figures on percentages of total production capacity for each of them are insignificant and cannot be calculated.

price reduction, of October 1984 and January 1985, respectively, were in operation at the time of writing, the chances of price stabilization were virtually nonexistent. Some of the most obvious reasons for that dim prognosis are as follows: (1) OPEC was producing about 49.8 percent of its total production capacity at the end of 1984 (Table 40), and at the conclusion of the first quarter of 1985 its production was not expected to be much higher. (2) At this rate, most oil states were producing at considerably below their maximum sustainable capacity. Saudi Arabia and Iraq—two of the three largest OPEC producers—were producing at far less than half, and Iran was producing at slightly over half of its maximum sustainable capacity. (3) Consequently, both high absorbers and low absorbers in OPEC reported substantial deficits in their current accounts (see Table A-8), and further production cutbacks appeared likely to put tremendous stress on their economies.

The greatest uncertainty faced by OPEC throughout its life has been a sustained inability to have a significant impact on the dynamics of the world oil market. In the buyers' market of the 1960s, the organization's performance was miserable. In the 1970s, despite a number of significant changes brought about by its members (elimination of the oil corporations as the major actors in setting prices, the conclusion of participation agreements, imposition of the oil embargo by OAPEC, and a series of unrealistic price increases in 1973 and again during 1979 and 1980), the effectiveness of OPEC remained subservient to temporary mutations in the oil market. Even though OPEC demonstrated its ability to *exploit* supply shortages, such as those stemming from the OAPEC-imposed embargo in 1973-1974 and from the political revolution in Iran, at no time was OPEC itself responsible for *creating* these shortages.

Throughout the 1970s OPEC proved its resiliency in the markets favoring the oil-exporting countries. If the first five years of the 1980s are an indication of things to come in the remainder of this decade, the world oil market promises to favor the buyers. Assuming that the survival of OPEC is vital to the prolonged economic prosperity of its members, these states have no choice but to utilize their diminishing oil power positively. Such a use of power may keep it from eroding to a point of no return. The greatest challenge to OPEC in the 1980s is to adapt to the new market conditions. A good starting point is to look at avenues leading to a resurgence in worldwide demand. About the only way to create such a resurgence is through a significant lowering of prices, possibly to about $22/b for heavy crudes and even lower for light crudes. At this scale the oil states may not have to absorb production cutbacks and their deleterious side effects. This price scale may also noticeably slow the adoption of conservation measures in the industrial countries. A substantial price rollback may also require an equally substantial reduction in the overly ambitious economic development projects adopted by the oil states in the 1970s. This in turn would substantially lower revenue needs, especially since the prices of almost all imported items show continual inflationary increases.

The oil states may also want to consider a substantial expansion of their downsteam activities, primarily by increasing and improving their refining capabilities. As Table 41 indicates, a number of OPEC members have already made noteworthy progress in this area. Venezuela and Saudi Arabia are clearly the leaders. Kuwait in recent years has been aggressively expanding its refineries. Iran and Iraq, two nations with considerable expansion potential, have been sidetracked by their protracted war. It might be argued that in the depressed market of the 1980s, refinery expansion may not be a profitable venture. But two students of oil affairs, Fereidun Fesharaki and David Isaak, go so far as to observe that the depressed market conditions and attendant overcapacity in the OECD refining industry are indeed the right circumstances for OPEC to enter the downstream end of the market. By absorbing financial losses in a depressed market and by continually improving their refining capabilities, these authors suggest, the OPEC nations may assure themselves a major role when the refining industry regains its strength.[5]

The political tensions involving Iran are a bomb ticking very close to the

Table 41. OPEC Refinery Capacity, 1982-1988 (1,000 b/d)

	Existing	Projects completed						Total projects	Total capacity
	1982	1983	1984	1985	1986	1987	1988	1983-88	1988
Algeria	438.0	—	—	—	—	—	—	—	438.0
Ecuador	94.5	—	15.0[1]	—	—	—	—	15.0	109.5
Gabon	44.0	—	—	—	—	—	—	—	44.0
Indonesia	471.0	360.0[2]	—	—	145.0[3]	—	—	505.0	976.0
Iran	560.0	—	—	—	—	—	200.0[4]	200.0	760.0
Iraq	305.5	150.0[5]	—	—	—	—	—	150.0	455.5
Kuwait	594.0	—	170.0[6]	—	206.[6,7]	—	—	376.0	970.0
Libya	130.0	220.0[8]	—	220.0[9]	—	—	—	440.0	570.0
Nigeria	247.0	—	—	—	—	—	—	—	247.0
Qatar	10.5	50.0[10]	—	—	—	—	—	50.0	50.0[10]
Saudi Arabia	1028.0	495.0[11]	513.0[12]	—	—	300.0[13]	—	1308.0	2336.0
UAE	135.0	60.0[14]	60.0[15]	—	—	—	—	120.0	255.0
Venezuela	1314.1	—	—	—	—	28.0[16]	—	28.0	1342.1
Total	5371.6	1335.0	758.0	220.0	351.0	328.0	200.0	3192.0	8553.1

Source: Abdallah al-Mangosh, "The Evolution of the Refining Industry," *OPEC Bulletin* 14, no. 6 (Aug. 1983): 18

[1]Expansion of Esmeraldas.
[2]Expansion of Cilacap and Balikpapan (year 1984, may be 2000).
[3]New refinery at River Musi.
[4]New refinery in Ilam Province planned.
[5]Baiji.
[6]Mina Al-Ahmadi expansion.
[7]Mina Abdullah expansion.
[8]Ras Lanuf.
[9]Misurata
[10]Umm Said (old unit 10.5 will be closed in 1984).
[11]Yanbu I (domestic, 170.0 + Rabigh 325.0).
[12]Yanbu II (expert 263.0 + Al-Jubail 250.0).
[13]Shuqaiq and Buraidah (150.0 each, both domestic).
[14]Umm Al-Nar expansion.
[15]Al-Ruwais expansion.
[16]Puerto La Cruz 25.0 and Cardon 3.0 expansion.

heart of OPEC. Considering the customary velocity with which political conflicts heat up and explode in the Middle East, these tensions have every potential for expanding from a regional conflict to a conflict involving the two superpowers. In the coming years, the survival of OPEC will be especially difficult because its members are likely to get little sympathy and even less help from the industrial consumers. Thus, in addition to all other uncertainties and challenges faced by this organization in the 1980s, OPEC has to survive on its own.

For a number of industrial states, the potential collapse of OPEC is not only a desirable scenario but also an applaudable one. Because a group of Third World oil states dared to hold the West "over a barrel"—through intermittent and unrealistic price increases—the conventional wisdom in the West is that every opportunity under the present market conditions to dismantle OPEC should be exploited. Such conventional wisdom is superficial and shortsighted, for it is oblivious to the political and strategic ramifications of such a happenstance. If one can get away from cyclical and fruitless arguments over the morality of price increases, it is a fact that oil revenues significantly improved the standard of living of people in a number of the oil states. A continued loss of these revenues because of depressed prices, together with the chaos that would be associated with the dismantlement of OPEC, promise to create political turmoil in an already turbulent Third World. Given the current realities of superpower politics, if the West will not respond to such chaos by intervening politically, economically, or militarily, the Soviet Union will. The political conditions prior to the Soviet intervention and occupation of Afghanistan continue to serve as an uneasy reminder to the West of its benign neglect of that unhappy country. The present revolutionary Iran appears to be edging toward increasing chaos and may become an awfully enticing case for Soviet adventurism. The potential collapse of OPEC may plunge the oil states of the Persian Gulf into similar political limbo.

Despite growing energy efficiency and fuel-switching, the non-Communist world remains enormously dependent on the use of oil. Just this fact alone forces the industrial West to make sure that access to oil supplies will not be jeopardized either as a result of regional turmoil or as an outcome of Soviet takeovers. The prudent strategy for the West is to allow OPEC to *adapt* to the market conditions of the 1980s. If OPEC is dismantled, let it be dismantled on its own, as a natural outcome of market forces, and not as a result of any political plots or conspiracies.

Appendix: Statistical Tables

Table A-1. Estimated OPEC Production Capacity, 1980

Producer	Million b/d
Saudi Arabia	12.0
Iran	4.0
Iraq	4.0
Kuwait	3.0
Venezuela	2.5
Nigeria	2.0
Libya	2.0
UAE	2.0
Indonesia	1.5
Algeria	1.0
Qatar, Gabon, Ecuador	1.0
Total	35.0

Source: Supplement to *Middle East Economic Survey,* Sept. 29, 1980.

Table A-2. Crude Oil Production by Iraq and Iran, 1979-1981 (1,000 b/d)

	Iraq	Iran
1979	3,477	3,168
1980		
Jan.	3,400	2,295
Feb.	3,400	2,500
March	3,400	2,350
April	3,300	2,200
May	3,300	1,700
June	3,300	1,500
July	3,100	1,700
Aug.	3,100	1,600
Sept.	3,000	1,400
Oct.	150	600
Nov.	350	800
Dec.	450	1,360
1980 Average	2,514	1,662
1981		
Jan.	600	1,600
Feb.	700	1,700
March	1,000	1,700
April	1,000	1,600
May	1,000	1,500
June	1,000	1,600
July	1,100	1,400

Source: EIA, *Monthly Energy Review*, Nov. 1981.

Table A-3. OPEC-wide Proved Oil Reserves and Production, End of 1981

Producer	Proved Oil Reserves, 1/1/82 (million barrels)	Estimated Oil Production, 1981	
		(1,000 b/d)	% of Change from 1980
Saudi Arabia	164,600	9,642	+0.1
Iraq	29,700	892	−64.5
Iran	57,000	1,375	−17.3
Kuwait	64,480	916	−34.0
Neutral Zone	6,500	370	−31.2
Libya	22,600	1,063	−40.5
Nigeria	16,500	1,369	−33.4
Algeria	8,080	750	−25.9
Venezuela	20,300	2,093	−3.4
Indonesia	9,800	1,607	+1.9
UAE			
Abu Dhabi	30,600	1,145	−15.2
Sharjah	306	9	−10.0
Dubai	1,270	358	+2.6
Qatar	3,434	414	−12.3
Ecuador	850	204	
Gabon	480	147	−16.0

Source: *Oil and Gas Journal*, Dec. 28, 1981

Table A-4. OPEC's Share of World Crude Oil Production, 1970-1981

Year	% of World	% of Non-Communist Areas
1970	50.6	61.2
1971	51.8	63.0
1972	52.7	64.1
1973	55.5	67.6
1974	54.9	68.0
1975	51.2	65.4
1976	53.5	68.2
1977	52.4	67.1
1978	49.5	64.2
1979	49.6	63.9
1980	45.2	59.5
1981	40.2	54.4
1982	35.1	48.5
1983	32.8	45.7
1984	32.1	44.4

Source: Percentages for 1971-1980 are from *Oil and Gas Journal*, Sept. 14, 1981. Percentages for 1981-1984 are calculated from data contained in *Oil and Gas Journal*, March 8, 1982, March 14, 1983, March 12, 1984, and March 11, 1985.

Table A-5. Saudi Arabian Crude Oil Production, 1979-June 1981 (1,000 b/d)

Month	1979	1980	1981
Jan.	9,790	9,785	10,265
Feb.	9,780	9,780	10,265
March	9,780	9,790	10,110
April	8,790	9,765	10,195
May	8,780	9,775	10,140
June	8,780	9,775	10,180
July	9,780	9,765	10,170
Aug.	9,770	9,765	10,330
Sept.	9,780	9,740	9,155
Oct.	9,725	10,255	9,685
Nov.	9,795	10,265	8,640
Dec.	9,530	10,260	8,645

Source: EIA, *Monthly Energy Review*, Dec. 1980, Oct. 1982.

Table A-6. Iranian Crude Oil Production and U.S. Imports from Iran, 1978-1980 (1,000 b/d)

Month	Iranian Crude Production	U.S. Imports from Iran
1978[a]		
Jan.	5,340	689.6
Feb.	5,580	539.2
March	5,650	535.2
April	5,660	441.9
May	5,770	746.3
June	5,680	536.0
July	5,850	532.5
Aug.	5,860	574.2
Sept.	6,100	590.6
Oct.	5,540	608.2
Nov.	3,540	494.7
Dec.	2,420	368.8
Average	5,240	555.3
1979		
Jan.	410	187
Feb.	760	86
March	2,190	22
April	3,800	52
May	4,100	197
June	3,950	318
July	3,750	425
Aug.	3,600	516
Sept.	3,600	373
Oct.	3,930	496
Nov.	3,170	549
Dec.	3,000	414
Average	3,168	304
1980		
Jan.	2,295	80
Feb.	2,500	9
March	2,350	—
April	2,200	—
May	1,700	—
June	1,500	—
July	1,700	—
Aug.	1,600	—
Sept.	1,400	—
Oct.	600	—
Nov.	800	—
Dec.	1,360	—
Average	1,662	

Source: EIA, *Monthly Energy Review,* July and Dec. 1980, Oct. 1981.

Table A-7. World Oil Production by Region, 1977-1984

	1977		1978		1979		1980	
Region	1,000 b/d	% of change	1,000 b/d	% of change	1,000 b/d	% of change	1,000 b/d	% of change
North America								
Canada	1,397	+7.2	1,324	+0.2	1,496	+13.9	1,412	−5.6
United States	8,225	+1.4	8,680	+6.1	8,598	−1.2	8,569	+0.3
Latin America	4,505	+3.9	4,757	+5.8	5,257	+10.5	5,586	+6.2
Europe	1,365	+59.1	1,748	+28.1	2,273	+30.0	2,473	+8.9
Africa	6,246	+7.5	6,132	−1.6	6,549	+4.1	6,032	−9.1
Middle East	22,152	+1.3	21,144	−4.5	21,005	+1.4	18,379	−14.5
Asia/Pacific	2,767	+9.9	2,797	+0.6	2,859	+2.2	2,732	−5.2
Communist areas	13,113	+5.3	13,755	+4.9	14,237	+2.3	14,488	+1.8
Total	59,769	+4.4	60,335	+1.2	62,767	+3.7	59,067	−5.0

	1981		1982		1983		1984	
Region	1,000 b/d	% of change	1,000 b/d	% of change	1,000 b/d	% of change	1,000 b/d	% of change
North America								
Canada	1,285	−10.9	1,241	−3.4	1,384	+11.5	1,434	+3.6
United States	8,580	—	8,656	+0.9	8,665	+0.2	8,759	+1.2
Latin America	5,971	+6.9	6,178	+3.5	6,072	−1.7	6,236	+2.7
Europe	2,568	+7.3	2,844	+10.7	3,302	+13.7	3,595	+8.9
Africa	4,541	−24.7	4,544	+0.1	4,333	−4.3	4,690	+7.5
Middle East	15,697	+14.6	12,383	−21.1	11,496	−6.8	11,406	−0.8
Asia/Pacific	2,779	+1.7	2,639	−6.2	2,843	+7.8	3,142	+13.7
Communist areas	14,596	+0.2	14,705	+0.7	14,923	+0.9	14,983	+0.4
Total	56,016	−6.2	53,191	−5.1	53,018	−0.5	54,245	+2.4

Source: *Oil and Gas Journal,* Feb. 27, 1978; Feb. 26, 1979; Feb. 25, 1980; March 9, 1981; May 8, 1982; March 14, 1983; March 12, 1984; March 11, 1985.

Table A-8. OPEC's Balance of Payments on Current Account, 1977-1985 (million $)

	1977	1978	1979	1980	1981	1982	1983	1984	1985	1986[c]
"Low absorbers"[a]	26	8	34	87	71	11	−2	10	8	6
"High absorbers"[b]	3	−10	31	23	−20	−26	−16	−8	−11	−12
Total	28	−1	65	111	51	−15	−18	2	−4	−6

Source: Extracted from *OECD Economic Outlook,* Dec. 1984, p. 150.
[a]Saudi Arabia, Kuwait, Libya, Qatar, Oman and UAE.
[b]Algeria, Ecuador, Gabon, Indonesia, Iran, Iraq, Nigeria, and Venezuela.
[c]Seasonally adjusted at annual rates.

Table A-9. Improvement in U.S. Energy Efficiency, 1929-1983 (all energy sources)

Year	GNP Constant 1972 $ (billion)	Energy Consumption (trillion BTUs)	Energy Consumption per GNP $ (thousand BTUs)
1929	315	23,756	75.4
1933	222	16,900	76.1
1940	343	23,589	68.8
1946	477	30,494	63.9
1950	534	33,922	63.5
1955	655	39,703	60.6
1960	737	44,569	60.5
1965	926	53,343	57.6
1970	1,075	67,143	62.5
1971	1,108	68,348	61.7
1972	1,171	71,643	61.2
1973	1,254	74,212	59.2
1974	1,246	72,479	58.2
1975	1,232	70,485	57.2
1976	1,298	74,297	57.2
1977	1,370	76,215	55.6
1978	1,439	78,039	54.2
1979	1,479	78,845	53.3
1980	1,475	75,900	51.5
1981	1,514	73,940	48.8
1982	1,485	70,822	47.7
1983	1,535	70,573	46.0
Growth rates:	%	%	%
1929-1940	+8.9	−0.7	−8.8
1940-1950	+55.7	+43.8	−7.7
1950-1970	+101.3	+97.9	−1.6
1970-1983	+42.8	+5.1	−26.6

Source: *Oil and Gas Journal,* July 30, 1984, p. 156.

Glossary

Affiliate A company whose relation to another company is basically similar to, but looser than, the relationship of parent and subsidiary. One company is affiliated with another if there is a bond of common ownership, directly or through or by another company, and if the operations of the two are subject to some common management or planning. Companies are not affiliated merely by virtue of a term supply contract.

Barrel (b or bbl) A liquid measure of oil, usually crude oil, equal to 42 American gallons or 280-380 pounds, depending upon API gravity, and equal to 35 Imperial (British) gallons.

Book Value The value of capital stock as indicated by the excess of assets over liabilities; or the value of equipment without regard to the actual or potential value of the product produced by the equipment.

British Thermal Unit (BTU) The quantity of heat necessary to raise the temperature of one pound of water one degree Fahrenheit. One BTU equals 252 calories, 1 gram (mean), 778 foot-pounds, 1,055 joules, and 0.293 watt-hours.

Buy-Back An agreement whereby the holder of an oil concession agrees to repurchase oil which, under the terms of a participation agreement, belongs to the producing country's government. The repurchase is usually made at or near posted price.

Cif Cost, insurance, and freight. Used to indicate that the price includes in a lump sum the cost of the goods, insurance, and freight to the intended destination.

Concession Any substantive right to explore for and develop petroleum and gas. A typical concession specifies areas to be explored, the duration of the concession, and terms of compensation to the host government.

Cracking Breaking down of large hydrocarbon molecules in heavy oils into a mixture of smaller molecules that can be distilled into gas, gasoline, distillate heating fuel, and other products.

Crude Oil Oil in its natural state, before refining or processing. It is a mixture of hydrocarbons that exists in natural underground reservoirs. It is liquid at atmospheric pressure after passing through surface separating processes. It does not include natural gas products, but does include the initial liquid hydrocarbons produced from tar sands, gilsonite, and oil shale.

Crude Oil Stocks Stocks held at refineries and at pipeline terminals, not including those held on leases (at storage facilities and adjacent to the wells).

Dead Weight Tons The total lifting capacity of a ship, expressed in long tons (2,240 lbs). For example, the oil tanker Universe Ireland is listed as 312,000 dwt, which means it can carry 312,000 tons of oil, or about 1.9 million barrels.

Desulfurization The process by which sulfur and sulfur compounds are removed from gases or liquid hydrocarbon mixtures.

Differential A premium added to the price of oil by the producing country to capitalize on certain advantages, such as low sulfur content or shorter hauling distance.

Distillation The first step in a refining operation in which crude oil is vaporized and then cooled and condensed to form "raw" gasoline, kerosene, heating oil, and other products. These raw products usually require further refining to meet quality specifications.

Downstream In a direction away from the wellhead and toward the gas pump; especially, refining and marketing activities.

Finished Products Petroleum oils, or a mixture or combination of such oils, or any components(s) of such oils, that are to be used without further processing.

Fossil fuel Any naturally occurring fuel of an organic nature, such as coal, crude oil, or natural gas.

fob Free on board. Used to indicate that the price includes loading aboard ship but not the cost of insurance or freight to the intended destination.

Independent In international oil, a company other than one of the international majors. In the U.S., usually one of the smaller, nonintegrated companies involved in only one phase of the oil industry, such as a local chain of gas stations or a fuel oil distributor.

Indexing A system whereby income adjustments are tied to inflationary increases in prices, interests, cost of living, and other transactions. Nations using indexation legally enforce all adjustments that are affected by inflation.

posted price An arbitrary value placed on a barrel of crude oil for the purpose of computing the amount of revenue the company must pay to the country. Posted prices may or may not approximate the market price or market value of the oil. A computation of revenues and company costs of a hypothetical posted price for a barrel of oil would be as follows:

Posted price	$4.00
Minus: production cost	.50
Minus: royalty (12½ percent of $4)	.50
Equals: tax reference price	3.00
The government would receive:	
Royalty (12½ percent of $4)	.50
Plus: tax (50 percent of tax reference price of $3)	1.50
Total revenue to government	2.00
The cost to the company would be:	
Production cost	.50
Plus: revenue to Government	2.00
Total cost to company or the tax paid cost	2.50

Probable reserves A realistic assessment of the reserves that will be recovered from known oil or gas fields based on the estimated ultimate size and reservoir characteristics of such fields.

Production programing (prorationing) A system for regulating oil and gas production to firm up prices by assigning the participating actors production quotas.

Production sharing An agreement between a company and the government of the country in which it is operating, under which the company gives part of its production (crude oil) to the government in lieu of cash.

recycling The process by which an oil-producer's investible funds are moved into final investments, either directly or through the intermediary of banks and institutions often located in a different country from the final destination of investment.

Refining The separation of crude oil into its component parts and the manufacture of products needed for the market. Important processes in refining are distillation, cracking, chemical treating, and solvent extraction.

Reserves Identified deposits known to be recoverable with current technology under present economic conditions. Categories of reserves include: (1) measured reserves, whose location, quality, and quantity are known from geologic evidence supported by engineering evidence: (2) indicated reserves, knowledge of which is based partly on specific measurements, samples, or production data, and partly on projections for a reasonable distance on geological evidence; (3) inferred reserves, based upon broad geologic knowledge for which quantitative measurements are not available; they are estimated to be recoverable in the future as a result of extensions, revisions of estimates, and deeper drilling in known fields; and (4) proved reserves, the small part of deposits known from exploration which has been readied for production by the drilling and equipping of wells; they represent only the amount of oil that can be produced at existing operating cost.

Royalties The owner's share of product or profit. Oil companies pay royalties to governments for the right to explore, for the right to pump oil, or for the right to put the oil on the market.

special drawing rights (SDRs) a form of liquidity reserves intended to increase total liquidity available for payments among members of the International Monetary Fund. SDRs augment gold, key international currencies, bilateral borrowing arrangements between nations, and borrowing facilities.

wellhead The location of an oil well. Also, oil or gas brought to the surface, ready for transportation to refinery or ship or pipeline. Wellhead costs usually include the cost of transporting, refining, or distributing the product for profit.

Sources: U.S. Congress, Senate, *Glossary of Terms Related to Petroleum*, prepared for the Subcommittee on Multinational Corporations of the Committee on Foreign Relations, 93 Cong., 2 sess. (Washington, D.C.: GPO, 1974); U.S. Congress, House, *Basic Energy Data and Glossary of Terms*, prepared for the Interstate and Foreign Commerce Committee, 94 Cong., 1 sess. (Washington, D.C.: GPO, 1975); *The McGraw-Hill Dictionary of Modern Economics*, 3rd ed. (New York: McGraw-Hill, 1983).

Chronology of Oil-Related Events, 1960 to March 1985

1960

Sept. 14 Organization of Petroleum Exporting Countries (OPEC)

1961

Jan. 21 Qatar joins OPEC. OPEC Conference approves statutes on structure of the organization and functions of its various bodies.

Nov. 1 OPEC expresses full support for Iraq in its dispute with concessionaire oil companies.

1962

June 8 Libya and Indonesia join OPEC. OPEC recommends that members demand posted prices not lower than pre-August 1960 prices and royalty payments to concessionaires at a uniform scale.

1965

July 13 OPEC Economic Commission recommends production programing.

1967

Sept. 17 OPEC supports Libyan effort to raise prices based on current freight rates.

Oct. 9 OPEC authorizes Saudi Arabia to negotiate with oil companies to eliminate allowances stipulated in expensing of royalty agreements for expensing of royalties.

Dec. 29 OPEC instructs its Economic Commission to continue search for economically practical system of production programing.

1968

Jan. 10 Iran, Kuwait, Libya, Qatar, and Saudi Arabia accept oil companies' offer on elimination of marketing allowance.

June 25 OPEC members intend to negotiate for "governmental participation" in ownership of concession holding companies.

Dec. 10 OPEC recommends that members "seek to ensure that the posted or tax reference prices . . . are consistent with each other, subject to differences in gravity, quality and geographic locations."

1969

July 9 OPEC secretary general directed to study linkage of posted prices to those of manufactured goods of major industrial countries.

Sept. 1 Libya's King Idris Sannusi overthrown in a bloodless coup. Revolutionary Council, headed by Muammar Qaddafi, announces it will honor existing agreements with oil companies.

1970

Jan. 16 Qaddafi becomes premier and defense minister of Libya, replacing Mahmoud Suleiman Maghraby.

Jan. 29 Qaddafi reportedly asks 10 percent increase in the $2.20/b posted price of Libyan oil.

April 8 Qaddafi threatens unilateral action if oil companies remain intransigent.

April 22 Libyan Oil Minister Izzedin al-Mabrouk rejects offers by Esso and Occidental.

June 12 Libya informs Occidental that total production of Libyan crude will be reduced from 800,000 to 500,000 b/d.

July 5 Libya announces nationalization of its four oil distributing companies.

July 9 Total production of Oasis reduced to 895,000 b/d, down 150,000 b/d.

July 27 OPEC supports Algerian demand to increase tax-reference price based on locational advantage.

Aug. 4 Production rate of Esso cut from 740,000 to 630,000 b/d.

Aug. 8 Posted price negotiations resume after reports of Libyan ultimatum. Occidental's offer of 20¢/b increase reportedly rejected.

Aug. 13 Esso reportedly offers 13-23¢/b increase.

Aug. 19 Libya again reduced Occidental's production to 440,000 b/d.

Sept. 4 Qaddafi signs agreement with Occidental raising posted price from $2.23 to $2.53/b.

Sept. 28 Esso and British Petroleum unilaterally and voluntarily raise posted price for Libyan crude from $2.23 to $2.53/b for Esso 40° API, and from $2.10 to $2.40/b for BP 37° API.

Dec. 28 OPEC announces intent to raise minimum level of income tax on oil firms 50-55 percent and to eliminate existing parities in posted price of crude on the basis of highest posted price applicable in member countries, taking into consideration differences in gravity and geographic location, and appropriate future escalation.

1971

Jan. 3 Libya tells oil companies that negotiations will begin Jan. 12 on a 5 percent increase in tax rate, and on a premium for Libya's freight advantage.

Jan. 12 OPEC members meet with 17 Western oil companies in Teheran to press demands for unspecified higher payments.

Jan. 16 In an effort to obtain an "overall and durable settlement," 15 Western oil companies propose to negotiate simultaneously with 10 OPEC members.

Jan. 24 Shah of Iran threatens oil cutoff if current talks collapse.

Feb. 2 Negotiations between 22 Western oil companies and Persian Gulf members of OPEC collapse.

Feb. 4 OPEC announces Feb. 15 as deadline for settlement; otherwise it will unilaterally raise prices.

Feb. 7 OPEC threatens "appropriate measures," including a total embargo, if oil companies reject higher prices.

1971 (Continued)

Feb. 14 Iran, Iraq, Saudi Arabia, Kuwait, Abu Dhabi, and Qatar reach five-year agreement with Western oil companies, including increases in payments of more than $1.2 billion for 1971. Agreements yet to be reached with four other oil-producing countries.

Feb. 23 Libya asks posted price increase of $1.02/b, total freight differential of 69¢, 5 percent increase in tax rates, and 25¢/b reinvestment tax.

April 2 Libya signs five-year agreement with companies, raising price from $2.55 to $3.45/b.

Aug. 15 President Nixon announces devaluation of U.S. dollar.

Sept. 22 OPEC formally demands participation negotiations.

Dec. 7 Libya nationalizes holdings of BP because of purported "conspiracy" between Britain and Iran in Iranian seizure of three islands in Strait of Hormuz.

1972

Jan. 15 Participation talks between six OPEC countries and companies open in Geneva. OPEC indicates that agreement must be reached before end of 1972.

Jan. 20 Six Persian Gulf OPEC members sign an agreement with Western oil companies in Geneva to compensate them for Dec. 1971 devaluation of U.S. dollar. Posted price raised 8.49 percent.

March 3 Saudi Arabia and other governments reject Aramco participation offer. Extraordinary OPEC Conference called.

March 10 On the eve of Extraordinary Conference, Aramco accepts notion of immediate 20 percent government participation.

March 17 OPEC agrees to majority takeover of Western oil companies if participation talks fail.

March 24 Iraq Petroleum Co. accepts principle of 25 percent participation in Iraq. Other companies accept 20 percent participation for Abu Dhabi, Qatar, and Kuwait.

June 1 Iraq nationalizes IPC.

June 28 Iran drops out of Persian Gulf group negotiating with companies. Shah announces separate deal with Iranian Consortium.

June 30 OPEC Conference in Vienna threatens "definite concerted action" if participation talks fail.

July 14 Saudi Arabia hints at some form of government takeover if participation talks fail.

Oct. 5 Agreement on participation finally reached in New York between major internationals and five Persian Gulf oil states represented by Yamani.

Oct. 9 Libya and ENI agree to 50-50 sharing of ENI's production in Libya, with ENI to market Libya's share for first five years.

Oct. 25 Persian Gulf states meet in Riyadh to iron out differences on participation agreements.

Dec. 20 Participation agreements with some revisions, formally signed by companies and governments of Saudi Arabia and Abu Dhabi.

1973

Feb. 13 U.S. announces second devaluation of dollar by 10 percent.

March 22 OPEC appoints team to renegotiate and amend "supplemental" to Teheran agreement of Dec. 20, 1972.

April 1 Nigeria acquires 35 percent interest in BP-Shell concession.

May 13 Qaddafi says Arabs will use oil as a self-defense weapon.

May 15 Kuwait, Libya, Iraq, and Algeria temporarily halt oil production to protest Israel's continued occupation of Arab territories.

May 16 Sadat urges Arab oil states to use oil to pressure U.S. into abandoning Israel.

June 1 Agreement announced between oil industry and OPEC to compensate for second devaluation of U.S. dollar.

Sept. 1 Libya nationalizes 51 percent of assets of all foreign oil companies.

Oct. 6 Hostilities break out between Arab and Israeli forces.

Oct. 15 U.S. announces it is resupplying Israel with military equipment.

Oct. 16 Six Persian Gulf states announce price increase of 170 percent.

Oct. 17 Organization of Arab Petroleum Exporting Countries (OAPEC) announces 5 percent monthly reduction to U.S. and other pro-Israel nations.

Oct. 18 Saudi Arabia announces 10 percent reduction in oil output to pressure U.S. to reduce support for Israel, and threatens to cut off all supplies to U.S. if aid to Israel continues.

Oct. 19 Nixon asks Congress for $2.2 billion for emergency military aid for Israel.

Oct. 20 Saudi Arabia announces halt in oil exports to U.S.

Oct. 21 Kuwait, Qatar, Bahrain, and Dubai announce total embargo of oil to U.S., theoretically completing cutoff of Arab oil. Iraq nationalizes Dutch oil holding in retaliation for Holland's pro-Israel leaning.

Oct. 26 Nixon chides European allies for withholding support for U.S. actions and policies in Middle East.

Nov. 1 Arab oil embargo spurs U.S. action on mandatory oil allocation program.

Nov. 5 France and Great Britain sidestep issue of pooling oil with European neighbors, including Netherlands, even though Dutch have been deprived of two-thirds of normal imports. Dutch and West Germans call for "solidarity" on this issue.

Nov. 6 European Economic Community urges Middle East war belligerents to return to Oct. 22 boundary lines.

Nov. 9 Secretary of State Kissinger fails to extract from King Faisal a promise of relaxation of Saudi oil embargo, but convinces king of his sincerity in search for Middle East peace.

Nov. 18 OAPEC, in a conciliatory gesture to Western Europe, announces cancellation of planned 5 percent production cut slated for December, but continues embargo to U.S. and Netherlands.

Nov. 21 Kissinger says U.S. will not alter its Middle East policies because of embargo and warns of countermeasures if embargo continues "unreasonably" and indefinitely.

Nov. 22 Yamani threatens to cut oil production by 80 percent if U.S., Europe, or Japan takes measures to counter embargo.

Nov. 30 OPEC declares 51 percent participation under agreed-upon schedule is "insufficient and unsatisfactory."

Dec. 16 Heads of nine EEC member governments agree, after long and bitter debate, to face oil crisis together, but only at price of a new statement on Middle East.

1973 (Continued)

Dec. 24 Six Persian Gulf oil states announce posted price of $11.65/b, effective Jan. 1.

Dec. 26 OAPEC announces cancellation of 5 percent production cut scheduled for January and promises 10 percent production increase, but continues embargo on supplies to Netherlands and U.S.

1974

Jan. 1 OPEC price increases announced Dec. 23 become effective.

Jan. 9 OPEC decides to freeze oil prices until Apr. 1.

Feb. 11 Libya nationalizes California Standard, Texaco, and Atlantic Richfield because companies will not agree to 51 percent participation.

Feb. 13 Saudi Arabia, Egypt, Algeria, and Syria decide to maintain oil embargo. In Washington, 13 industrialized countries agree to form coordinating body to counter oil price increases.

Feb. 25 Coordinating committee formed at Washington energy conference meets to prepare for oil consumers-producers meeting in May.

March 13 Consumers coordinating committee meets in Brussels to decide future priorities; oil consumers-producers meeting rescheduled for late June.

March 18 OAPEC announces suspension until July 1 of oil embargo against U.S.

May 17 EEC releases its study on revitalizing European coal industry.

May 23 France announces plans to order 48-55 new nuclear power plants by 1980.

May 31 Syrian-Israeli disengagement agreement reduces possibility that OAPEC embargo against U.S. will be reinstated after June 30 deadline.

June 2 OAPEC announces that embargo against U.S. will not be reinstated at end of June.

June 5 Saudi Arabia and Aramco agree to increase government participation in company from 25 percent to 60 percent, retroactive to Jan. 1, 1974.

June 10 EEC agrees unanimously to cooperate with Arab nations in economic and technical affairs.

June 17 In Quito, OPEC conference decides to freeze prices through third quarter of 1974, to raise royalty payments from 12.5 to 14.5 percent (about 10¢/b, and to hold current production levels.

July 9 In Brussels, 12-nation Energy Coordinating Group (France refuses to attend) agrees to share energy resources in times of crisis, but postpones proposed meeting with oil-producing states.

July 10 OAPEC ends embargo against Netherlands and approves studies on alternative energy sources and nonfuel use of petroleum.

Sept. 4 Iran says it will propose replacement of posted/fiscal/royalty pricing system with set price tied to worldwide inflation rate.

Sept. 13 In Vienna, OPEC decides to maintain posted price freeze through fourth quarter of 1974 but to increase royalty from 14.5 to 16.67 percent and tax rate from 55 to 65.75 percent, resulting in effective increase in government take of 3.5 percent. OPEC spokesmen say increases should be absorbed by companies, not passed on to consumers.

Sept. 16 EEC members initiate agreement for coordinated energy strategy, including cooperation in research, conservation, and resource sharing in time of shortage.

Sept. 18 Norway reportedly reconsidering its position on Energy Coordinating Group's proposal to share energy supplies.

Oct. 15 Iran reportedly will propose replacement of posted price system with fixed price of $9.84/b, linked to prices of twenty to thirty other world commodities as hedge against inflation.

Nov. 1 French President Giscard d'Estaing joins Shah of Iran and President Boumedienne of Algeria in calling for international meeting to discuss oil prices and international economic problems.

Nov. 9 Saudi Arabia, Qatar, and Abu Dhabi announce posted price reduction of about 40¢/b and increase in royalty and tax rates, changes that will raise price paid for equity crude but lower price of participation oil, giving state-owned producers an advantage in open market.

Nov. 19 In Paris, 16 of OECD's 24 members agree to join International Energy Agency, which will promote cooperation in energy conservation, research and development, nuclear enrichment, and investments, and International Energy Program for sharing resources in times of shortage.

Dec. 1 Yamani says Arabs will be absolute owners of their natural resources before end of year, implying they will either purchase or nationalize remaining assets of foreign oil companies.

Dec. 9 Saudi Arabia and Aramco agree in principle to 100 percent government participation and open negotiations on Saudi compensation for remaining 40 percent of Aramco's assets. Qatar, on Dec. 23, and UAE, on Dec. 27, announce they will begin similar negotiations.

Dec. 12 In Vienna, OPEC replaces posted price system with fixed market price of $10.12/b for Arabian Light marker crude, effective Jan. 1, 1975, through Sept. 30, 1975.

Dec. 30 Kissinger says U.S. will consider "indexing" oil to other commodities if there is an initial drop in oil prices.

Dec. 31 Libya ends oil embargo against U.S.

1975

Jan. 8 Kissinger does not rule out use of military force to seize oil fields if "industrialized world" is being economically strangled.

Jan. 14 Secretary of Defense Schlesinger says it is militarily feasible to invade oil-producing countries but should be done only in gravest emergency.

Jan. 27 OPEC endorses proposal for an international energy conference of oil-producing states, industrialized nations, and developing countries.

Feb. 3 Kissinger says oil-consuming nations could break OPEC cartel in two years if they could agree to "a common floor price" for imported oil at a level that would stimulate conservation and development of alternative energy sources.

Feb. 7 IEA approves energy conservation plan to save 2 million b/d in 1975, and agrees to preliminary meeting with oil-producing states.

Feb. 27 OPEC oil ministers say they will revise oil prices upward in three months if dollar continues to decline.

March 5 Kuwait announces it will nationalize remaining 40 percent of oil industry, currently held by Gulf and BP.

1975 (Continued)

March 6 At summit meeting, OPEC agrees to meet with oil-consuming and developing nations on Apr. 7 to discuss oil price freeze, linking price of oil to world inflation, protection of producers' investments in West, reform of world monetary system, aid to developing nations, and measures to counter threats or aggression against oil fields.

March 17 Ninth Arab Petroleum Congress in Dubai denounces oil-consuming nations for trying to blame oil producers for world energy problems, and reaffirms producers' right to use oil as a weapon against Israel.

March 21 IEA meeting in Paris ends with agreement in principle to establish a floor price for oil and to cooperate in developing alternative energy sources.

March 25 OPEC meets to discuss how to cope with declining dollar. Iran, Saudi Arabia, Kuwait, and Qatar have severed links with dollar, tying their currencies to Special Drawing Rights of IMF; Venezuela is considering such a move.

April 1 Economic Commission of OPEC discusses substituting Special Drawing Rights for dollar and establishing production quotas to maintain higher prices.

April 9 Industrialized nations formally establish "safety net" fund of $25 billion to help financially troubled countries pay for oil imports.

April 15 Paris meeting of oil-consuming, oil-producing, and developing nations, which began April 7, breaks down because of disagreement over agenda and participation in full meeting proposed for late summer 1975. U.S. favors agenda limited to energy, while oil-producing states favor inclusion of all raw materials, Third-World development, and energy.

June 12 In Gabon, OPEC nations agree to postpone until Sept. a decision to change from dollar to IMF's SDRs. Delegates also discuss price increase slated for Sept. and possible resumption of producer-consumer meeting, and approve full membership of Gabon in OPEC. Britain's first North Sea oil well begins production; full production expected in 1980.

July 4 IEA reportedly agrees to resumption of oil producer-consumer dialogue suspended Apr. 15, and accepts agenda desired by oil producers.

July 28 IEA agrees to raise individual countries' oil stocks from 60 to 70 days.

Sept. 27 OPEC announces price increase of 10 percent, from $10.46 to $11.51/b from Oct. 1 to June 30, 1976. OPEC ministers postpone until Dec. decision on switching from dollar to SDRs.

Dec. 1 Kuwait agrees to pay $50.5 million for remaining 40 percent interests of Gulf and BP in Kuwait oil industry.

Dec. 19 At end of 3-day meeting in Paris, Conference on International Economic Cooperation passes resolutions to establish commissions on energy (co-chairmen Saudi Arabia and U.S.), raw materials (Japan and Peru), development (Algeria and EEC), and finance (EEC and Iran).

1976

Jan. 12 Industrial, raw materials-producing, and oil-producing states begin informal meetings to map strategies for four CIEC commission meetings, to begin in Feb.

Jan. 28 OPEC agrees to establish an $800-million fund to assist poor nations.

March 5 OECD reports that OPEC aid to developing countries totalled $5.6 billion for 1975, compared to $4.6 billion for 1974.

March 12 Aramco and Saudi Arabia announce "general accord" on sale of Aramco's remaining 40 percent of operations in Saudi Arabia.

April 9 U.S. Energy Administrator Zarb predicts possibility of an even worse oil embargo and says OPEC not likely to be broken up, despite U.S. actions.

May 10 OPEC finance ministers discuss establishing a long-term, no-interest loan fund for poor nations and agree to contribute $400 million to a $1 billion agricultural fund.

May 29 OPEC members cannot agree on price increase and decide to maintain price freeze indefinitely.

June 14 Several OPEC members reduce price of heavy crude oil by 5-10¢/b following decision in May to adjust price differentials between heavy and light crudes. Libya and Algeria raise price of their light crudes; Venezuela, acting contrary to OPEC decision, raises price of its heavy crude.

July 10 CIEC Energy Commission ends first phase of meeting to analyze international energy problems. Second phase, recommending solutions, to be completed before CIEC Ministerial Conference scheduled for Dec. 1976.

Aug. 28 OPEC economic ministers reportedly cannot agree on a common formula for price differentials, due to be considered at next OPEC meeting in Dec.

Sept. 9 U.S. FEA announces choice of 8 underground sites for storage of 150-million-barrel strategic oil reserve to counteract effects of future oil embargoes. Reserve to be in place by 1978.

Sept. 24 Reports from Persian Gulf show spot prices for crude oil running about 20¢/b above long-term contract prices, reinforcing producers' claims that prices will be raised in Dec.

Oct. 1 At end of CIEC Energy Commission meeting, EEC proposes permanent energy consultative forum to continue discussions after CIEC completes work in Dec. IEA report on energy conservation shows 19 member states have reduced 1975 energy consumption 4.8 percent below 1973 rate, and chastises U.S. for wasting energy.

Oct. 15 Reports from Persian Gulf and Caribbean indicate spot market prices for crude continue to creep upward in anticipation of OPEC price rise in Dec.

Nov. 22 Shah says Iran willing to arrange a bilateral indexing system to peg oil price to other trade commodities.

Dec. 8 CIEC meeting scheduled for Dec. 15 postponed to Mar. 1977.

Dec. 17 Eleven members of OPEC announce price increase of 10 percent, effective July 1, 1977; Saudi Arabia and UAE announce increase of 5 percent effective Jan. 1. Later, Indonesia drops announced increase from 10 to 6 percent.

Dec. 21 Yamani says Saudi Arabia will not increase production past 8.5 million b/d ceiling already established, countering a rumor that Saudis would flood market to drive world oil prices down.

1977

Jan. 1 Price increases on crude oil announced by OPEC on Dec. 17, 1976, go into effect. Saudi Arabia and UAE increase prices 5 percent; Indonesia, 6 percent; other ten members, 10 percent.

Jan. 3 U.S. Department of Commerce estimates that U.S. imported 2.6 billion barrels of oil in 1976 for $32 billion, compared to 2.2 billion barrels for $25 billion in

1977 (Continued)

1975. Imported oil for 1977 is expected to cost $2-4 billion more than for 1976.

Jan. 6 Saudi government and Aramco officials announce that lower price of Saudi crude will be passed on to consumers because no brokers will be allowed to handle Aramco oil, companies will have to file audit reports of their costs, and new production will be allocated at same rate as old. Statement includes veiled threat to cut off any company that does not comply.

Feb. 28 OPEC finance ministers meet in Vienna to distribute $800 million in interest-free loans to needy Third World countries that have been hurt by OPEC price increases.

March 1 Reports from OPEC headquarters suggest that members that raised prices by 10 percent on Jan. 1 will forgo additional 5 percent raise scheduled for July 1 if Saudi Arabia and UAE raise prices to 10 percent level.

April 18 OECD reports that U.S. 1976 oil imports rose 25 percent over 1975, despite government pledges to reduce imports.

May 11 Explosion and fire at Abqaiq, Saudi Arabia, destroys a pumping station and pipeline complex, interrupting flow of about 6 million b/d. Oil is rerouted within a week.

June 1 CIEC participants agree on need for conservation, new resource development, and international cooperation in energy affairs. Participants do not agree on price of energy resources, recycling "petrodollars," financial assistance for oil-importing developing nations, or form for future international energy consultations.

June 4 A second fire breaks out at Abqaiq, Saudi Arabia, but is extinguished quickly. Saudi government discounts rumors of sabotage.

June 22 Negotiations are reported completed for Saudi purchase of remaining 40 percent of Aramco's holdings.

July 12 OPEC meeting at Saltsjobaden, Sweden, primarily concerned with production programing to protect price levels.

Sept. 6 Kuwait lowers price of its heavy crude by 12¢/b to counter oversupply of heavy crude on world market.

Oct. 5 Governing board of IEA adopts 12-point policy guideline which includes cutting oil consumption, reducing oil imports, developing alternative energy resources, and increasing use of nuclear energy.

Nov. 28 Prime Minister Amouzegar says Iran will support a two-year price freeze, reinforcing statement by Shah that Iran will not push for price increase at Caracas.

Dec. 21 In Caracas, OPEC members agree to six-month price freeze because they cannot agree on how much to raise prices. Venezuela proposes 5-8 percent increase, with additional proceeds given to Third World countries to help retire their $180 billion debt.

1978

Jan. 16 In a visit to Saudi Arabia, U.S. Energy Secretary Schlesinger reportedly convinces Saudis to continue accepting U.S. dollars as payment for oil in spite of dollar's falling value.

Jan. 22 Tenth Arab Petroleum Congress meeting in Tripoli, Libya, calls for indexing oil prices, more production control, more refineries, more research and development efforts on alternative energy sources, and more economic cooperation, and reasserts need to "deploy Arab oil in the service of Arab causes."

Feb. 1 Five producers of heavy crudes (Venezuela, Iran, Saudi Arabia, Iraq, and Kuwait) agree to decrease in price of Kuwaiti crude to protect its market position.

April 10 Leaders of national oil companies of OPEC countries agree to twice-yearly meetings to discuss cooperation in recovery, downstream operations, and other company problems and to coordinate relations with foreign contractors.

May 7 OPEC oil ministers meet to discuss prices and production controls. They agree to freeze prices for remainder of 1978.

June 19 In Geneva, OPEC agrees to maintain price freeze through end of year, but oil ministers predict 1-2 percent increase in early 1979. In Europe, reports say recent six-week test of IEA emergency allocation system was successful. Next test scheduled for end of 1979.

July 14 OPEC meets to discuss declining value of dollar and possibility of shifting to a basket of currencies for pricing oil.

July 17 At seven-nation economic summit in Bonn, President Carter says U.S. will lower oil imports and raise coal production to reduce U.S. trade deficit and stabilize dollar. U.S. Congress considers legislation to ban oil import fees, one method believed under consideration by president.

Nov. 24 Ayatollah Khomeini, leader of religious opposition to Shah of Iran, calls for a new oil workers' strike to begin Dec. 2, and for destruction of Iran's oil installations if government opposes strike.

Nov. 26 Pemex officials say Mexico's proven oil reserves are 20.2 billion barrels, probable reserves are 37 billion, and potential reserves are 200 billion and current 880,000 b/d production will be raised to 2.25 million by 1980.

Dec. 18 OPEC announces it will raise oil prices by annual rate of 14.5 percent, beginning with 5 percent increase on Jan. 1, and further increases on Apr. 1, July 1, and Oct. 1, 1979.

Dec. 20 Except for a "trickle" of some 400,000 b/d, Iranian oil industry is shut down by oil worker strikes and public disturbances.

1979

Jan. 1 Oil production in Iran is completely shut down by strikes and continuing political crisis. Several thousand foreign oil technicians began leaving in last week of Dec., and by early Jan. only a handful remain.

Jan. 17 Schlesinger says U.S. oil shortage of about 500,000 b/d caused by Iranian shutdown is "serious" but not yet "critical." Other producers make up about 300,000 b/d of 800,000 b/d in crude oil and refined products lost.

Feb. 17 Iranian oil workers end strike. National Iran Oil Co. says first exports will leave Iran Mar. 5 and production will be up to 3 million b/d within two weeks.

Feb. 19 In Paris, IEA Executive Director Ulf Lantzke says Iranian shutdown has caused an overall drop in supplies of less than 4 percent, not enough to trigger automatic emergency sharing system. He also says member-state stockpiles are

1979 (Continued)

"marginally higher" than in Jan. 1978, suggesting IEA members are better prepared to withstand oil shortage.

Feb. 22 Libya raises oil price by 5 percent, following similar increases by Venezuela, Kuwait, Abu Dhabi, and Qatar.

Feb. 28 Hassan Nazih, head of National Iranian Oil Co. (NIOC) says Iran will no longer honor consortium arrangement and will deal directly with companies offering to purchase oil.

March 3 An NIOC spokesman says Iran's production target is 4 million b/d.

March 7 NIOC announces production has reached 2 million b/d, and export target is 3 million b/d.

March 13 Exxon announces it will stop selling crude to non-affiliated companies in 1981.

March 19 Mexico announces it will raise prices to keep pace with other producers. On Apr. 1, Mexican crude goes to $17.10/b.

March 26 At OPEC meeting in Geneva, Iraq advocates price rise of 24 percent to keep up with inflation; Saudis advocate holding to 3.8 percent price increase scheduled for Apr. 1. OPEC compromises by raising prices to 14.5 percent scheduled for Oct., an increase of 9.5 percent for Apr. 1.

April 10 Iranian oil production reaches 4.4 million b/d and an export level of about 3.5 million b/d.

April 14 U.S. Senate Committee on Foreign Relations releases report that says Saudi Arabia will not raise production above 12 million b/d in late 1980's, as past Saudi statements have said.

May 16 Yamani says Saudi Arabia will maintain production levels above its ceiling for rest of year to support U.S. peace efforts, but if U.S. fails to arrange a comprehensive peace, they will cut back production.

May 22 IEA governing board reaffirmed its intention to reduce oil consumption by 5 percent, or 2 million b/d. IEA recognizes that economic growth cannot be sustained without resolution of energy problems and that oil cannot meet medium-term energy needs in future.

May 31 During past month, Oman, Ecuador, Iran, Abu Dhabi, Libya, and Venezuela have raised prices to $17/b for marker crude. OPEC reportedly will raise prices in June, 20-35 percent.

June 7 NIOC head Hassan Nazih says Iranian oil production is holding steady at about 4 million b/d with exports at about 3.2 million b/d, despite political disruptions, and announces increase to $18.47 for Iranian Light and $17.74 for Heavy. Iran is accepting no new oil customers because it has enough revenue.

June 22 EEC agrees to freeze oil imports at 470 million tons per year through 1985.

June 28 OPEC raises Saudi Light from $14.54 to $18/b and places a ceiling of $23.50/b on higher grade crudes. Most other OPEC members add surcharges to raise their comparable crudes to $20/b.

June 29 Italy, France, West Germany, Britain, Canada, Japan, and U.S. agree to set limits on oil imports through 1985. EEC members sought a 5-year overall ceiling, but U.S. and Japan held out for two-year individual country ceilings.

July 5 OECD reports that U.S. oil consumption decreased 0.7 percent in first quarter of 1979 while consumption in Japan and Western Europe increased.

Sept. 28 IEA study says U.S. and Canada should increase gasoline and fuel oil prices to encourage conservation and urge Japanese to increase conservation efforts. It lists three leading oil users (tons of crude oil per capita per year): Luxembourg, 10.77; Canada, 8.7; U.S., 8.28. For comparison, West Germany uses 4.25 and Japan 2.4.

Oct. 3 Mana Said Utaiba, UAE oil minister, tells OPEC energy seminar that inflation and declining dollar will lead to another price increase.

Oct. 4 Iran reportedly raises price of spot sales of crude to Japanese companies to $40/b; Japanese see this as a step toward a major increase at Dec. OPEC meeting and a hint that Iran will not offer long-term supply contracts.

Nov. 7 Kuwait announces it will reduce contract sales by 500,000 b/d, confirming rumors that Kuwait will reduce production from 2 million to 1.5 million b/d.

Nov. 14 President Carter freezes Iranian assets in American banks in response to seizure of American embassy in Teheran on November 4. Oil-producing states protest U.S. move.

Dec. 3 Following meeting of OPEC strategy committee (Kuwait, Saudi Arabia, Iraq, Iran, and Venezuela), Yamani announces OPEC may replace dollar as medium of exchange for international oil transactions with a basket of currencies to protect against fluctuations.

Dec. 10 IEA votes to reduce total oil imports of its 20 members to 24.6 million b/d by 1985, from 1979 ceiling of 26 million b/d. Guido Brunner, EEC oil commissioner, tells IEA that EEC will support U.S. and refrain from purchasing Iranian oil on spot market.

Dec. 13 Saudi Arabia, UAE, Qatar, and Venezuela raise oil prices by $6/b to forestall a larger increase at Dec. OPEC meeting. Saudi marker crude is set at $24/b.

Dec. 17 At OPEC meeting, Libya announces price increase of $4.75/b, to $30; Iran raises its prices by $5/b, and Indonesia by $4.40, an apparent response to Saudi, Venezuelan, Qatari, and UAE increases announced Dec. 13.

Dec. 20 OPEC meeting ends without agreement on prices for coming year. Apparently, each member state will set its own prices according to market, as in previous six months. Many oil analysts suggest such open pricing will create a glut, which will drive prices down.

1980

Jan. 28 Saudi Arabia raises its oil by $2/b, (to $26/b for Arabian Light), in line with comparable crudes from Venezuela, Kuwait, Iraq, Indonesia, Ecuador, UAE, and Qatar. The next day, Kuwait, Iraq, Qatar, and UAE raise prices by $2/b.

Feb. 22 OPEC strategy committee (Saudi Arabia, Iraq, Venezuela, Iran, Kuwait, and Algeria) discusses holding four rather than two pricing meetings each year and considers basket of twelve currencies to replace dollar.

April 21 U.S. Energy Secretary Charles Duncan says U.S. will not resume filling strategic reserve until June 1981. U.S. last purchased oil for reserve in Nov. 1978; it has 91 million of projected 750 million barrels. Most oil-producing states have said they will not sell oil for stockpiling.

1980 (Continued)

May 14 Saudi Arabia raises its benchmark crude from $26 to $28/b, apparently to forestall a larger increase, rumored at $5/b, at OPEC's June meeting. Within one week, Nigeria, Venezuela, Qatar, Kuwait, and UAE raise prices by $2/b. Angola and Oman, not OPEC members, raise prices on May 24 and 25, respectively.

June 11 Eleven OPEC ministers agree to raise Saudi Light by $2, to $32/b. Saudi Arabia and UAE hold at $28/b. Kuwait immediately announces an increase from $29.50 to $31.50/b. The $32 benchmark price will have an upper limit of $37 for higher grade African crudes. There is no decision on production restrictions or ways to stop stockpiling.

June 26 Seven industrialized nations blame high rates of inflation and unemployment on OPEC pricing policies. OPEC Secretary General Rene Ortiz blames the seven for world inflation through their poor administration, irrational energy consumption, failure to fund alternative energy resources, and overpricing of exports.

July 1 Libya, Nigeria, Kuwait, and Algeria announce immediate increases of approximately $2/b (variable according to quality from $31.50 to $37.00/b). Iraq, Venezuela, and Mexico follow shortly with similar increases.

July 14 Saudi Foreign Minister Saud al-Faisal says his country will maintain production at 9.5 million b/d for rest of 1980. Announcement is expected to maintain downward pressure on oil prices. Current spot market prices are equal to or below, OPEC prices.

Aug. 26 UAE announces it will lower oil production to keep prices higher, and will advocate similar moves by other members of OPEC at Sept. 15 meeting. Saudi Arabia proposes that OPEC review prices quarterly and adjust for inflation, economic growth rates, and currency fluctuations.

Sept. 9 Reports say UAE is willing to raise its oil price by $4/b and that Saudis are contemplating a production cutback of about 1 million b/d to unify prices and reduce crude glut. Indonesians reportedly advocate a unified rather than staggered price, which currently ranges from $28 (Saudi Arabia) to $37 (Algeria).

Sept. 15 At OPEC ministers' meeting, Iran, Algeria, and Libya refuse to discuss Saudi plan to link oil prices to inflation, currency, and economic growth fluctuations. Iran asks for a change of venue for 20th anniversary meeting scheduled for Nov. in Baghdad, because of Iraq-Iran border tensions.

Sept. 17 OPEC members agree to lower price of Saudi Light marker crude from $32 to $30/b, and Saudis agree to raise 34° API crude from $28 to $30/b, temporarily creating a unified price. (Announcement is seen as gesture toward Saudis.) Rumors circulating after meeting say members also agreed to reduce production by 10 percent to eliminate world glut and protect prices.

Sept. 21 Algeria announces it will reduce production by 100,000 b/d, about 10 percent. Next day, Libya, Iraq, and UAE also announce 10-percent reductions.

Sept. 23 Several OPEC members, led by Venezuela and UAE, say they will forgo 10-percent reduction to cover anticipated shortages.

Sept. 25 Iran and Iraq begin bombing each other's oil facilities as their border war expands. Iranian refinery at Abadan, world's largest, is reportedly burning and badly damaged, and Iranians retaliate against Iraqi refinery complex at Basra. Some 40 oil tankers and freighters are trapped in 100-mile long Shatt al-Arab waterway. Oil tanker traffic through Strait of Hormuz is slowed as shipowners delay moving in or out of Persian Gulf for fear of being caught in conflict.

Sept. 30 Ulf Lantzke, executive director of IEA, says loss of nearly 4 million b/d from Iran and Iraq will not cause a serious shortage for IEA members before end of year.

Oct. 1 OPEC finance ministers' meeting, scheduled for Oct. 6-7 in Quito, postponed at request of Iran and Iraq because of war. Ministers are in Washington for World Bank and IMF meetings. In Kuwait, OAPEC announces cancellation of Rome meeting to discuss southern Europe-OAPEC development cooperation.

Oct. 2 Saudi Arabia, UAE, Kuwait, Qatar, and Indonesia reportedly planning to raise production to compensate for shortfall caused by Iran-Iraq war. Sources disagree on amount of Saudi increase (ranging from 600,000 to 1.5 million b/d.) Other producers expected to increase about 100,000 b/d each.

Oct. 9 Kuwait announces it will not increase oil production beyond current 1.5 million b/d.

Oct. 10 President Shahu Shagari says Nigeria will not increase oil production to compensate for Iran-Iraq shortfall because it would cause a drop in prices. Next day, Mexico says it will not increase oil production.

Oct. 21 IEA's governing board meets to discuss oil supplies in wake of Iran/Iraq war, sees no oil supply problems through first quarter of 1981. By coincidence, IEA was finishing a two-month test of its oil-sharing plan when war broke out.

Oct. 26 UAE raises oil price from $31.56 to $33.56/b in line with Saudi $2 price increase of Sept. Oil Minister al-Utaybah says increase has nothing to do with shortfalls caused by Iran/Iraq war. Kuwait denies it will raise its price by $5/b.

Oct. 29 Rotterdam spot market hits $40/b, up about $3/b since start of Iran/Iraq war and up $8/b over a month before.

Oct. 30 Saudi Arabia reportedly planning to limit production to 10 million b/d rather than the 10.4 million b/d agreed to.

Nov. 2 Kuwait repeats Oct. 9 statement that it will not increase production above current 1.5 million b/d. Next day, UAE announces it will hold production to 1.8 million b/d.

Nov. 18 Iraq notifies long-term contract oil customers that some oil (estimated at 6-10 million b/d) will be available at Mediterranean end of Iraq-Turkey pipeline beginning Nov. 20, the first Iraqi exports since late Sept.

Dec. 16 OPEC announces a 10 percent price increase: Saudi Light marker, the industry standard, from $30 to $32/b, and Libyan and Algerian high-quality lights from $37 to $41/b. Several non-OPEC producers raise prices as well: Great Britain (North Sea oil) by $2/b; Mexico, by $4/b, effective Dec. 24; Egypt by $4.50/b; Oman by $6/b; China, by $2.66/b.

Dec. 30 *Oil and Gas Journal* reports that world oil production fell 5 percent, to 59.6 million b/d, for 1980. U.S. increased production slightly, as did Soviet Union and China; Middle East production declined because of Iran/Iraq war. World oil

1980 (Continued)

reserves increased by 1.1 percent (7 billion barrels), to 648 billion barrels. World gas reserves increased 2.5 percent, to 2.6 quadrillion cubic feet.

1981

Jan. 12 *Petroleum Intelligence Weekly* reports that the 21 IEA nations cut 1980 oil imports by 10 percent and oil consumption by 6 percent. IEA target for 1981 is to hold imports at present levels and increase consumption by 0.3 percent. *Mideast Report* says Iran is exporting about 250,000 b/d from Persian Gulf ports of Lavan Island and Bandar Mah Shah. Iran's primary Gulf loading terminal at Kharg Island was destroyed in early days of Iran/Iraq War.

Jan. 23 Nigerians raising price of crude by $3/b rather than $4/b as called for by Libya and Algeria, other OPEC members producing oil similar to Nigeria's.

March 2 *Petroleum Intelligence Weekly* reports that Saudi Arabia will reduce oil production by 500,000 b/d in late March to tighten market and support higher prices. Current Saudi production is 10.1 million b/d. Kuwait will reportedly lower production ceiling from 1.5 million to 1.2 million b/d.

March 17 Oil ministers of Saudi Arabia, Kuwait, Qatar, and UAE meet in Riyadh to discuss growing oil glut and reduced pressures for high prices. Oil supplies have increased with resumption of Iraqi and Iranian production and with conservation efforts by industrialized countries.

March 18 Yamani denies Saudi Arabia will reduce production by 500,000 to 600,000 b/d. Saudi production remains at 10.1 to 10.3 million b/d.

March 30 Yamani says Saudi Arabia will maintain production at 10.3 million b/d until there is a unified OPEC oil price and OPEC agreement on long-term production strategy.

April 20 Yamani tells American television interviewer that Saudi Arabia engineered world oil glut by raising production after Iran/Iraq war erupted and is maintaining it to drive prices down.

May 25 Yamani hints that Saudis may raise price to $34/b, more in line with OPEC price of $36/b.

May 26 OPEC announces a freeze on current OPEC price of $36/b for Saudi Light and a ceiling of $41 for North African crudes for rest of year. A majority of OPEC members agree.

June 8 *Wall Street Journal* reports that Nigeria has lowered its crude from $40 to $36/b to regain customers. Nigerian production has fallen to 1 million b/d because of contract cancellations.

June 15 U.S. Energy Secretary James Edwards tells reporters at IEA Paris meeting that members should maintain reserves for emergencies and that self-triggering mechanisms for emergencies may need revising. Smaller members of IEA have been drawing upon emergency stockpiles rather than pay dollars for oil.

June 17 OPEC Long-Term Strategy Committee meets in Geneva to formulate a proposal on prices, production levels, and indexing of prices with world inflation levels.

June 25 Kuwaiti Oil Minister Sheikh Ali Khalifa al-Sabah says Kuwait may drop its surcharge, reported to be $3 or $4/b, but will not cut prices below OPEC standard in spite of glut.

June 26 *MEES* reports that Saudi Arabia will not reduce oil production as rumored. "War relief" oil (450,000 b/d produced to meet shortages caused by Iran/Iraq war in Sept. 1980) will be phased out after July 1, but Saudis will not reduce production levels below 450,000 b/d.

July 29 Venezuelan crude production drops another 40,000 b/d, to 1.8 million b/d, in line with its decision at May OPEC meeting to reduce production by 10 percent. Last year, Venezuela produced 2.2 million b/d, of which about 1.8 million b/d was exported.

June 30 Libya drops price of crude by $1/b, to $39.90, in line with comparable Nigerian and Algerian crudes at $40/b. Oil companies operating in Libya have sought a drop to $36/b, in line with Norwegian and British North Sea crudes. Contract negotiations between Libyan government and oil companies continue, despite June 30 expiration.

July 6 UAE Oil Minister Otaiba says OPEC should have a unified price and should address questions of price and supply. Otaiba says there is a 4 million b/d surplus on world oil market.

July 27 African members of OPEC say they will not lower prices to clear oversupplied market. Lighter grade African crudes usually sell about $5/b higher than lower-grade Persian Gulf or South American crudes.

Aug. 2 *MEES* reports OPEC members are moving toward Saudi position on price unification and production drops. Nigeria has agreed that multiple prices and wide spread between low- and high-grade crudes ($32-$43) is undermining OPEC position, a change from Nigerian position of July 27.

Aug. 18 Yamani says Saudi Arabia will not raise its crude price to OPEC level of $36/b but may go to $34; he agrees to drop current premiums for lighter-grade crudes and to freeze prices for an extended period. OPEC ministers begin preliminary meetings prior to OPEC pricing conference. Venezuela's representative stands by $36/b price.

Aug. 20 At OPEC pricing meeting in Geneva, Iraq offers a compromise of $35/b to unify prices, but Saudi Arabia stays with $34/b. Iran will not drop below $36/b nor agree to a long-term price freeze.

Aug. 21 OPEC pricing meeting ends without apparent agreement on pricing policies for coming year. Yamani says Saudi Arabia will cut production by 1 million b/d and will continue to sell at $32 for remainder of year.

Oct. 1 Kuwait claims three Iranian planes bombed oil storage center at Umm al-Aysh, destroying part of complex.

Oct. 29 OPEC agrees to unify price at $34/b and to freeze prices through end of 1982. Yamani syas Saudi Arabia will drop production to 8.5 million b/d on Nov. 1 and that price freeze will hold through 1982. He predicts oil glut will dry up by middle of 1982.

Oct. 30 EEC energy ministers agree to prevent price explosions in spot market during shortages by asking oil companies to avoid buying on spot market, to draw on commercial stocks rather than seek new purchases, and to replace oil with other energy resources. Move is designed to deal with shortages that are less than IEA definition of a crisis that would trigger IEA sharing scheme.

Nov. 9 Algeria and Libya lower price of high-grade crude by 50¢/b, to $37.50, in spite

1981 (Continued)

of their position at Oct. OPEC meeting. Differential now set at $3.50. Saudi sources discount rumors that they intend to build a billion-barrel strategic stockpile for crude oil at Yanbu, but do want to increase present storage capacity from 11 million to about 50 million barrels. Trans-Arabian pipeline, with an ultimate capacity of 3.7 million b/d, now handles about 1.5 million b/d; current plans call for an increase to 1.85 million b/d by mid-1982.

Dec. 7 Iran is reportedly averaging about 90,000 b/d in exports and 450,000 b/d for domestic use. Iranian Ambassador to Kuwait Ali Shams Ardakani says Iran will increase exports to 1.1 million b/d, and is receiving enough in oil revenue to meet its expenses, including war with Iraq.

Dec. 9 At OPEC meeting, Yamani says Saudi Arabia will maintain $34/b price through 1982, regardless of possible dollar devaluations or market glut. He predicts glut will end in spring 1982, while others predict it will last longer.

Dec. 11 In Abu Dhabi, OPEC members agree to lower price of heavy crudes by 50 cents to around $3/b, effective Jan. 1, 1982. Under new pricing arrangement, high-grade crudes from Nigeria, Algeria, and Libya will fall to around $36.50/b because of lower differentials for grade and distance, and Kuwaiti, Iranian, Saudi, and some UAE heavy crudes will fall below $34 Arabian Light price fixed at Oct. 1981 OPEC meeting. IEA governing board proposes that members supply IEA with monthly statistics on supply, demand, and price of oil as a way to stop "panic buying" during short-term supply emergencies. IEA believes that lack of information about market conditions led to "panic" spot-market buying which in turn led to 1979 price increase.

1982

Feb. 1 World oil markets appear poised for a new round of price cuts as weak demand for everything from crude oil to heating fuels continues in U.S., Western Europe, and Japan, despite an extremely cold winter.

Feb. 8 Britain and Iran cut oil prices, Iran by $1/b and Britain by $1.50/b; other cuts are possible.

March 3 OPEC officials begin preliminary talks and acknowledge that the alliance faces serious threat. Plunging oil sales are pressing Nigeria to break ranks with OPEC and cut price. Saudi Arabia steps up its pressure on oil companies to boost purchases of Nigerian oil and protect fragile OPEC pricing agreement. Britain, succumbing to world oversupply of petroleum, slashes price of North Sea crude by $4/b.

March 4 OPEC members call for a meeting following British oil price reduction.

March 11 Nigeria has hinted to petroleum companies it may cut oil prices as much as $5/b if OPEC does not act to curb oil glut.

April 8 Saddled with long-term oil contracts with countries such as Saudi Arabia and Nigeria, whose oil is considered overpriced, France moves to cut its losses without provoking political fury.

May 20 OPEC ministers indicate organization may try to keep industrialized countries from building substantial petroleum reserves.

June 21 Strains over production controls are reportedly having repercussions in some member countries.

July 8 OPEC calls emergency meeting for tomorrow to cope with member states that have been producing oil above their OPEC-mandated quotas.

July 12 OPEC disarray may spark cut in crude oil prices; failure to reach an accord on quotas and production mirrors group hostilities.

Sept. 15 Saudis to hold oil price at $34/b despite slump; Yamani says price cut would cause a shortage in world supply of crude.

Oct. 21 OPEC confronts deep split in ranks; Kuwait warns that a "significant drop" in oil prices will result if some OPEC members continue discounting and violating production ceilings.

Oct. 13 OPEC oil ministers drop plans for a Consultative Meeting in Vienna on Oct. 28 to settle differences over price and production timetables.

Oct. 14 Pressure eases on oil producers to reduce prices; escalation of Iran/Iraq war and approach of winter are reviving demand.

Nov. 24 OPEC faces stormier times ahead; Arab Gulf faction asks to delay Dec. 9 parlays; split over price policy.

Nov. 29 OPEC has pushed back its year-end meeting by 10 days, to Dec. 19, to give members more time to compromise their differences on oil prices and production.

Dec. 6 Independent study group concludes that oil prices may "collapse" by next spring unless OPEC acts immediately to redistribute production quotas among members and to end price discounting by petroleum producers.

Dec. 7 Yamani vows to defend OPEC's current price structure and to uphold OPEC benchmark price at $34/b.

Dec. 14 OPEC to focus on cutting oil production to boost prices, but outlook is uncertain.

Dec. 20 OPEC oil ministers launch major effort to avoid a price war and put an end to disagreements on production and pricing. They may boost Iran's quota at Saudi expense.

1983

Jan. 17 OPEC to meet in Geneva to try to keep prices from falling further; move comes after several members threaten to slash quotas.

Jan. 20 OPEC nears accord to trim oil production to avert price collapse; Saudi policy shift is spurred by report that U.S. plans rise in taxes if price falls; remaining snag is Venezuela.

Jan. 24 OPEC agrees to limit production but leaves benchmark price at $34/b; tentative accord hinges on some adjustments in differentials.

Jan. 25 OPEC meeting fails to agree on oil prices and output, clearing way for a possible round of lower quotas; Saudi Arabia and Gulf allies wanted other producers to boost charges.

Jan. 27 OPEC's oil pact collapses at last minute over price discounts; other members claim Saudis lured them into a trap to assert their position.

Feb. 2 After OPEC's failure to agree on production quotas, oil buyers and sellers are not asking *if* price of oil will fall, but when and how much.

Feb. 11 Saudi Arabia warns other oil producers that it is no longer committed to defending official crude price; unless other OPEC members agree to reduce their price, the kingdom will move alone to cut quotas.

1983 (Continued)

Feb. 18 Nigeria, having fallen on hard times, may be first in OPEC to cut oil prices.

Feb. 22-23 OPEC oil ministers meet amid indications that they plan to cut prices at least $4/b.

March 3 OPEC oil ministers will meet for latest in a series of mini-conferences in hope of agreeing on lowered oil prices and organizing a new production program. A concession toward Nigeria by Arab Persian Gulf producers moves OPEC closer to production and pricing agreement.

March 11 Oil ministers reach a "general understanding" on price reduction; price of Arabian Light marker crude to be cut $4-$5/b.

March 14 OPEC announces it will lower its marker by $5 to $29/b; sets daily output ceiling at 17.5 million b/d.

March 16 Soviet Union's need for Western currency could undercut OPEC's attempt to halt slide in oil prices; USSR may step up oil exports to compensate for lost revenue.

April 25 OPEC has told Algeria to negotiate with Soviet Union in further effort to coordinate oil pricing.

June 8 OPEC succeeds in returning a measure of stability to world oil markets despite minor violations of OPEC accords, which forbid granting discounts or exceeding production quotas.

June 9 Soviets emerge as a world oil power; fearing price decline, OPEC asks Moscow to restrain its output.

Sept. 16 OPEC decides against changes in its production ceiling of 17.5 million b/d; it will also retain current prices.

Sept. 30 Research group on OPEC policies cites new dangers ahead, such as pressure on oil prices, flat demand, and violation of production quotas by members.

Oct. 11 Iran again warns Arab oil exporters that it will halt oil shipments out of Persian Gulf if Iraq unleashes its French Super Etender warplanes; spot oil prices edge up in anticipation.

Oct. 27 OPEC is trying to come to grips with alleged production cheating in Saudi Arabia, Iran, and Nigeria that is depressing oil prices on world markets.

Nov. 17 Dispute among OPEC oil ministers over production and pricing policies over next few years pits Iran against Saudi Arabia.

Dec. 6 OPEC chiefs try to avert further oil price erosion, but weak demand and internecine fighting may take their toll; some fear that escalation of Iran–Iraq war could lead to disruption of oil shipping lanes.

Dec. 9 OPEC ministers will maintain a previous agreement mandating total oil production for members at 17.5 million b/d and a marker price of $29/b.

1984

Jan. 27 Soviet Union continues to make things difficult for OPEC by increasing oil shipments to West by 15 percent in past year.

Feb. 10 New budget in U.S. calls for more than 20 percent cutback in buildup rate for Strategic Petroleum Reserve; lawmakers of both parties up in arms, urge more purchases while prices are soft.

Feb. 14 Energy consumption among Common Market nations declined in 1983 for fourth year, but could rise slightly in 1984; oil consumption fell in 1983 while nuclear power and natural gas use rose.

Feb. 23 Iran attacks on Iraq worry U.S. and Britain; allies planning protection for oil supply route.

March 6 IEA predicts oil consumption in North America will rise 12 percent in first quarter from year before; predicts a 5.7 percent first-quarter rise in oil consumption for OECD as a whole, but oil consumption in Europe is expected to decline as much as 3 percent in the quarter.

March 7 Energy experts say OPEC stands to benefit by wave of U.S. oil business. Among other reasons, enormous amounts of money being gobbled up by mergers diminish capital available for finding new oil in U.S. and increase reliance on imports.

March 12 OPEC's Market Monitoring Committee says current worldwide balance of oil demand and supply does not warrant changes in group's production levels and prices.

March 24 Iran, concerned that rising Persian Gulf insurance rates will frighten ships away from its ports, is expanding an insurance program of its own to entice wary shippers; Saudi Arabia warns that a blockade of Persian Gulf could result in out-of-control surge in world oil prices.

April 6 Reagan Administration advocates increased federal intervention in energy marketplace in case of major emergency; Energy Secretary Hodel says that in an emergency, U.S. would begin immediately to sell crude oil in Strategic Petroleum Reserves, regardless of allies' steps to cope with shortage.

May 7 OPEC Market Monitoring Committee predicts that demand for OPEC oil will grow later this year, but advises against immediate changes in organization's prices or production quotas.

May 8 Soviet Union produced as much crude oil in first quarter as in year-earlier period, indicating that its oil-production problems may be easing.

May 17 Iran–Iraq war is moving dangerously close to Saudi Arabia's oil fields, raising risk of wider war in Gulf that could trigger U.S. military involvement; U.S. is ready to send help if Saudis request it; oil market gets more jittery in response to fears war will escalate.

May 18 Oil tanker traffic in Persian Gulf has dropped only slightly, despite recent attacks on ships by Iran and Iraq that have caused about $100 million in damages.

June 7 Iraqi and Iranian attacks on ships in Persian Gulf have had no significant effect on world oil supplies or prices, according to IEA.

June 12 Oil ministers of six Gulf Corporation Council countries say they will replace oil shipments lost due to air attacks in Persian Gulf to keep insurance rates on ships from rising; but insurance underwriters at Lloyds of London, reacting to confirmed attack on Kuwaiti tanker, double insurance rates on cargoes in southern part of Gulf.

June 13 Kuwait Petroleum Corp. is using its ships to carry crude oil from Persian Gulf to Gulf of Oman for loading onto tankers chartered to Far East and Europe; Kuwait's oil sales have been cut recently because of rising costs for ships sailing to its ports in Persian Gulf.

1984 (Continued)

June 14 An offer by Persian Gulf oil states to compensate ship owners for petroleum cargoes lost because of Iran–Iraq war, combined with discounts being granted by Iran and some other producers, appear to have dissipated fears among Gulf oil buyers.

June 20 Instead of raising oil prices, fighting in Persian Gulf is indirectly putting downward pressure on them; pressure is coming in short run from increased output by major producing countries which helps depress prices on spot market.

July 9 Several OPEC members arrive in Vienna for Ministerial Meeting expecting to keep prices and output unchanged and hoping to convince other members to stop cheating on production quotas.

July 12 OPEC, faced with desperate plea and veiled threat from Nigeria, raises its production quota by 150,000 b/d by Sept.

July 26 Nigeria says it will not give up its increased oil production quota when allotment expires at end of Sept., a move likely to aggravate OPEC's effort to keep oil prices from sagging further.

July 30 Soviet Union cuts oil price by $1.50/b; Urals brand of crude oil slashed to $27.50/b for Aug.

Aug. 6 OPEC crude oil production in July was 18.6 million b/d, well above second quarter average.

Aug. 15 Global production of oil increases 7 percent in 1984 first half from a year earlier, according to industry estimate; oil output from all producers rose to average of 54.7 million b/d in first six months of 1984.

Aug. 21 Oil production by OPEC is dropping below cartel's mandated ceiling of 17.5 million b/d.

Aug. 29 Yamani says OPEC may consider increase in oil production during fourth quarter of year.

Sept. 1 Soviet Union raises price of Urals crude oil export to Western Europe by 25¢/b to $28/b effective Oct. 1.

Sept. 6 U.S. dependence on imported oil likely to grow by 1990s despite conservation efforts, according to Office of Technology Assessment; result will be increased oil shortages worldwide.

Sept. 26 Nigeria and Iran have been offering lower prices for higher-grade crude oils for past month to spur sales.

Ot. 17 British National Oil Corp. postpones decision on whether to match Norway's price cuts for North Sea crude.

Oct. 18 In a move likely to trigger worldwide price cuts, Britain reduces crude oil prices by $1.35/b to average of $28.65/b; move comes in wake of recent declines in spot prices that have made it difficult for Britain and other producers to maintain official price quotes; OPEC expected to meet soon to decide how to react to price cut.

Oct. 19 Nigeria cuts crude oil prices by $2/b in major new downward push on world prices; OPEC announces emergency meeting for Oct. 29 in Geneva; OPEC leaders maintain they can prevent a general price decline.

Oct. 22-23 Seven OPEC ministers meet to avert a collapse of oil prices, a week ahead of emergency meeting of all oil ministers. They fail to produce a concrete proposal for holding oil prices at current levels; optimistic statement by Saudi minister helps push some energy prices higher.

Oct. 23 Oil-producing, debt-burdened Third World countries won't be hurt seriusly even if crude price cuts initiated by Norway, Britain, and Nigeria spread industry-wide; but economists warn that if price cuts deepen, several petroleum producers could be squeezed, among them Nigeria, Mexico, Venezuela, and Indonesia.

Oct. 26 Yamani predicts Nigeria will rescind its recent price cut, but refuses to say whether he has received such a commitment from that country.

Oct. 30-31 OPEC, in an effort to shore up crumbling oil prices, will lower its official production ceiling to 16 million b/d, a reduction of 1.5 million b/d; but cartel fails to agree on how to apportion cuts among members.

Nov. 1 OPEC says eleven of its thirteen members will reduce production ceilings to bolster prices.

Dec. 18 OPEC ministers begin discussions in Vienna on ways to stabilize oil prices amid speculation that there is little they can do to prevent further price drops next year.

Dec. 31 Algeria and Nigeria refuse to endorse internal pricing pact reached by OPEC.

1985

Jan. 9 BNOC confirms it is selling oil at spot-prices, but says this is not a change in pricing policy for North Sea oil.

Jan. 15 Norway and Britain quietly move closer to linking oil prices to spot market.

Jan. 22 OPEC's output in early Jan. is estimated at 15.5 to 15.9 million b/d, about 1 million b/d less than last quarter's average.

Jan. 31 For the second time in its history, OPEC cuts prices, and abandons $29/b benchmark. Algeria, Libya, and Iran reject this decision; Gabon abstains.

Feb. 4 Egypt lowers price of its main export crude. More reductions anticipated by non-OPEC members following cut by cartel.

Feb. 6 Mexico lowers prices of its oil exports following OPEC's lead.

Feb. 26 Dutch accounting firm KMG Klynveld Kraayenhoff is hired by OPEC to monitor member states' oil prices and production figures.

March 14 Britain to end North Sea oil price support and abolish BNOC this spring.

March 19 New slump in world oil demand seen following 1984 period of revival.

Sources: *New York Times Index*, 1960-1973; *Wall Street Journal Index*, 1982-1984; *Chronologies of Major Developments in Selected Areas of International Relations*, 1974-1976, 1978-1981; U.S. Congress, Senate, Subcommittee on Multinational Corporations of the Committee on Foreign Relations, *Multinational Corporations and United States Foreign Policy*, 93 Cong., 1 and 2 sess. (Washington, D.C.: GPO, 1974), pp. 4 and 5.

Notes

Chapter 1. Introduction

1. As concessionaires, the oil companies were granted rights to conduct various oil-related activities over a specified period. In return, the host governments received financial payments and royalties.

2. G. Gray, "Dangers in Worldwide Stagnation," *Euromoney,* Feb. 1979, pp. 49-50.

3. C.Y. Lyn and Abdul K.M. Siddiqui, "Recent Patterns of Inflation in Non-Oil-Producing Countries," *Finance and Development,* Dec. 1978, pp. 28-31.

4. According to one study conducted by the Department of Energy, "Energy prices made a significant contribution to the economic slowdowns of the last decade, but were not the dominant causes." U.S. Department of Energy, *Interrelationship of Energy and the Economy* (Washington, D.C., July 1981), pp. 1-2.

5. The non-oil-producing countries of the Third World were the worst victims of the OPEC price escalations of 1973. The adaptive capability and new cooperative arrangements between the Western industrialized countries seem to have ameliorated their suffering as they recuperated from the initial impact of the sellers' market during 1973 and 1974.

6. Charles Issawi and Mohammed Yeganeh, *The Economics of Middle Eastern Oil* (New York: Praeger, 1962), p. 22. For a comprehensive account of the activities of these companies, also see Harvey O'Conner, *World Crisis in Oil* (New York: Monthly Review Press, 1962), pt. 5; and George W. Stocking, *Middle East Oil: A Study in Political and Economic Controversy* (Nashville: Vanderbilt Univ. Press, 1970).

7. Zuhayr Mikdashi, *The Community of Oil Exporting Countries* (Ithaca: Cornell Univ. Press, 1972), p. 35.

8. Michael Tanzer, *The Political Economy of International Oil and the Underdeveloped Countries* (Boston: Beacon Press, 1970), p. 30.

9. Stocking, *Middle East Oil,* p. 8.

10. Ibid., p. 125.

11. Ibid., pp. 9-11.

12. For details, see Kamal S. Sayegh, *Oil and Arab Regional Development* (New York: Praeger, 1968), pp. 31-32; also see George Lenczowski, *Oil and State in the Middle East* (Ithaca: Cornell Univ. Press, 1960), chapt. 4.

13. The Anglo-Persian Company was formed in 1909 to take over the D'Arcy concession, which was revised to cover a total of 100,000 square miles. APOC changed its name to Anglo-Iranian Oil Company when Persia officially changed its name to Iran in 1935. In 1954, AIOC once again changed its name to British Petroleum.

14. O'Connor, *World Crisis in Oil,* p. 290.
15. Ibid., p. 286.
16. This discussion of nationalization of oil in Iran is based on O'Connor, *World Crisis in Oil*; Stocking, *Middle East Oil*; and Tanzer, *Political Economy of International Oil.*
17. For details, see David Wise and Thomas B. Ross, *The Invisible Government* (New York: Random House, 1964).
18. Theodore Moran's testimony, U.S. Congress, Senate, Subcommittee of the Committee on Foreign Relations, Hearings, *Multinational Corporations and the United States Foreign Policy,* 93 Cong., 1 and 2 sess. (Washington, D.C.: GPO, 1974), pt. 3, pp. 294-305. Hereafter cited as Senate, *Multinational Corporations.*
19. Fuad Rouhani, *A History of O.P.E.C.* (New York: Praeger, 1971), pp. 39-40.
20. Ibid., pp. 41-42.
21. The bulk of this analysis is based on U.S. Congress, Senate, *The International Petroleum Cartel,* staff report to the Federal Trade Commission, submitted to the Subcommittee on Monopoly of the Select Committee on Small Business, 82 Cong., 2 sess. (Washington, D.C.: GPO, 1952).
22. Ibid., pp. 200-201.
23. Edith Penrose, *The Large International Firm in Developing Countries: The International Petroleum Industry* (London: George Allen and Unwin, 1968), pp. 34-35 and 46.
24. Franklin Tugwell, "Petroleum Policy in Venezuela: Lessons in the Politics of Dependence Management," *Studies in Comparative International Development,* Spring 1974, pp. 86-87.
25. Ibid., pp. 86-88. According to Fadhil al-Chalabi, an OPEC official, before the establishment of OPEC, the decision on how much oil to produce in each oil state was made by the multinational corporations entirely on the basis of "the size and capacities of the global distribution networks for the final consumers on a world scale." The revenue needs or desires of the oil-producing countries to fully develop their oil potential were totally disregarded. Fadhil al-Chalabi, *OPEC and the International Oil Industry* (London: Oxford Univ. Press, 1980), pp. 12-16, *passim.*
26. Tugwell, "Petroleum Policy in Venezuela," p. 88. This does not rule out the possibility of common interests between the oil companies and the host government. Numerous examples of such congruence can be found. For instance, in the days before the oil embargo, executives of a number of major oil corporations lobbied for an evenhandedness in American policy in the Middle East, a position promoted by various oil states. When Congress was deliberating the possible sale of Airborne Warning and Control System (AWACs) to Saudi Arabia, numerous oil executives, along with Saudi officials, lobbied for passage of this legislation.

Chapter 2. The Uncertainties of the Buyers' Market

1. Fuad Rouhani, *A History of OPEC* (New York: Praeger, 1971), p. 45.
2. Ibid., p. 76.
3. OPEC, *Information Booklet* (Vienna: Jan. 1973), p. 5.
4. Zuhayr Mikdashi, *The Community of Oil Exporting Countries: A Study in Governmental Cooperation* (Ithaca: Cornell Univ. Press, 1972), p. 23.
5. Ibid., p. 24.

6. Prorationing refers to production control; it may also be called "production programing" and "planned production." See Glossary.

7. *Oil Imports and Energy Security: An Analysis of the Current Situation and Future Prospects,* report of the Ad Hoc Committee on the Domestic and International Monetary Effect of Energy and Other Natural Resource Pricing of the Committee on Banking and Currency, U.S. Congress, House, 93 Cong., 2 sess. (Washington, D.C.: GPO, 1974), p. 44.

8. Ibid.

9. Zuhayr Mikdashi, *A Financial Analysis of the Middle East Oil Concessions, 1901-65* (New York: Praeger, 1966), p. 172.

10. For a detailed analysis of this point, see Penrose, *Large International Firm in Developing Countries,* pp. 77-80.

11. See Glossary.

12. *International Petroleum Industry,* p. 192.

13. Benjamin Shwadran, *The Middle East, Oil and the Great Powers* (New York: John Wiley, 1973), p. 500.

14. Ibid.

15. OPEC, *Information Booklet,* p. 5.

16. Stocking, *Middle East Oil,* p. 352.

17. OPEC, *Information Booklet,* p. 5.

18. OPEC, Resolution I.1. (Roman numerals indicate the conference number; Arabic numerals indicate the resolution number.) *OPEC Official Resolutions and Press Releases, 1960-1980* (New York: Pergamon Press, 1980). All OPEC resolutions cited are from this source.

19. For further details, see Shwadran, *Middle East, Oil and the Great Powers,* pp. 506-7.

20. Abbas al-Nasrawi, "Collective Bargaining Power in OPEC," *Journal of World Trade Law,* March/April 1973, pp. 188-207.

21. Mikdashi, *Community of Oil Exporting Countries,* p. 48.

22. Similar views were expressed by Ashraf T. Lutfi, former secretary general of OPEC, in his book *OPEC Oil* (Beirut: Middle East Research and Publishing Center, 1968), p. 35.

23. Marketing expense was about 12.5% of the posted price. The oil companies deducted it from gross income before determining taxable income. This was considered the host countries' contribution toward marketing expense.

24. Rouhani, *History of OPEC,* p. 212.

25. Ibid., p. 214.

26. OPEC, Resolution II.13.

27. Franklin Tugwell, *The Politics of Oil in Venezuela* (Stanford: Stanford Univ. Press, 1975), p. 63.

28. Stocking, *Middle East Oil,* p. 383.

29. Rouhani, *History of OPEC,* p. 201.

30. Stocking, *Middle East Oil,* pp. 383-84.

31. OPEC, Resolution IX.61.

32. "OPEC's Recipe," *Petroleum Press Service,* Sept. 1965, pp. 328-31.

33. *Petroleum Intelligence Weekly* (hereafter cited as *PIW*), Aug. 30, 1965. *Supplement to Middle East Economic Survey* (hereafter cited as *MEES*), Feb. 11, 1966.

34. *PIW,* May 17, 1966. The Shah of Iran, in a speech in March 1966, and the Iranian prime minister, A.A. Hoveydya, in a statement before the Iranian parliament, vowed to

proceed with increased production. This was a blatant violation of prorationing. Also see *MEES*, March 22, 1968, and Dec. 18, 1968, "Supplement."

35. OPEC, Resolutions XIV.84 (passed in Jan. 1968), XX.112 (passed in June 1970), and XXI.121 (passed in Dec. 1970).

36. Rouhani, *History of OPEC*, p. 220.

37. Stocking, *Middle East Oil*, pp. 359-62.

38. Ibid., p. 359.

39. Shwadran, *Middle East Oil and the Great Powers*, pp. 160-61.

40. Ibid., p. 161.

41. Rouhani, *History of OPEC*, p. 229.

42. J.E. Hartshorn, *Politics and World Oil Economics* (New York: Praeger, 1962), pp. 342-43.

43. OPEC, *1968 Review and Record* (Vienna, 1969), pp. 4-5.

44. OPEC, Resolution VII.50.

45. Mikdashi, *Community of Oil Exporting Countries*, p. 116.

46. Resolution I.2; also see Resolution I.1.

47. Mikdashi, *Community of Oil Exporting Countries*, p. 130.

48. *Economist*, Aug. 10, 1963.

49. Kamal Sayegh, *Oil and Arab Regional Development* (New York: Praeger, 1968), pp. 167-68.

Chapter 3. The Making of the Sellers' Market

1. Regaei Mallakh, "The Suez Canal Closure: Its Costs for the United States," *Middle East Forum*, Autumn/Winter 1971, pp. 47-59.

2. The distance between the Persian Gulf states and the Western European markets is only 6,500 miles when the Suez Canal is open, a twenty-day one-way voyage, compared to a thirty-two-day trip when it is closed.

3. For a detailed account of Libya's humiliating and financially disastrous experience with the multinational oil companies, see Leonard Mosley, *Power Play: Oil in the Middle East* (New York: Random House, 1973), especially chap. 24, pp. 320-33.

4. Libyan demands were as follows:

- Libyan crude oils are fully eligible for a gravity differential of 2¢/bbl/degree (paraffinic crude might be an exception):
- Libyan crude oils have been denied a freight advantage of about 8-10¢/bbl when compared to Eastern Mediterranean postings;
- Libyan oil should be entitled to a Suez premium greater than what has been indirectly involved in the temporary elimination of allowances related to the royalty-expensing agreement:
- A sulphur premium for Libyan crude oils should be recognized in two respects: (1) savings in refinery maintenance costs; and (2) savings in sulphur pollution-abatement;
- A price increase should be implemented retroactively.

Taki Rifai, *The Pricing of Crude Oil: Economic and Strategic Guidelines for an International Energy Policy* (New York: Praeger, 1974), p. 252.

5. The bulk of the following discussion is based on U.S. Congress, Senate, Subcommittee on Multinational Corporations of the Committee on Foreign Relations,

Chronology of the Libyan Oil Negotiations, 93 Cong., 2 sess. (Washington, D.C.: GPO, 1974), pp. 4-12.

6. Libya wanted an increase of 40-50¢/b and a graduated scale of 2¢/API degree, a freight differential, and a credit for the low sulphur content. *MEES,* March 1, 1970.

In his testimony before the Senate subcommittee on multinational corporations, Piercy of Exxon stated: "The posted price dispute in Libya involved two basic issues—the quality differential accorded Libyan crude as compared to crudes in the Persian Gulf and the place, or freight, differential that Libyan crudes should be given because they are much closer to the European market than Persian Gulf crudes." Senate, *Multinational Corporations,* pt. 5, p. 214.

7. Marwan Iskandar, *The Arab Oil Question* (Beirut, 1973), pp. 17-18.

8. *Middle East Economic Digest* (hereafter cited as *MEED*), June 5 and Sept. 4, 1970.

9. Joel Darmstader and Sam Schurr, "The World Energy Outlook to the Mid-1980's: The Effects of an Alternative Supply Path in the United States," *Philosophical Transaction of the Royal Society of London,* Series A, May 1974, pp. 413-30.

10. Peter R. Odell, *Oil and World Power* (Baltimore: Penguin Books, 1970), p. 71.

11. Sam H. Schurr and Paul T. Homan, *Middle Eastern Oil and the Western World* (New York: Elsevier, 1971), p. 9.

12. *The Middle East: U.S. Policy, Israel, Oil and the Arabs* (Washington, D.C.: Congressional Quarterly Publication, April 1974), p. 26.

13. U.S. Congress, Senate, Committee on Foreign Relations, Hearings, *Energy and Foreign Policy,* 93 Cong., 1 sess. (Washington, D.C.: GPO, 1973), pp. 187-88.

14. U.S. Congress, House, Subcommittee on the Near East and South Asia of the Foreign Affairs Committee, Hearings, *Oil Negotiations, OPEC and the Stability of Supply,* 93 Cong., 1 sess. (Washington, D.C.: GPO, 1973), p. 3.

15. Senate, *Multinational Corporations,* pt. 5, p. 212.

16. Senate, *Chronology of the Libyan Oil Negotiations,* p. 6.

17. Ibid., p. 10.

18. Piercy's testimony in Senate, *Multinational Corporations,* pt. 5, p. 214. Another step-by-step description of events leading up to the Tripoli I agreement stated that "The government's move was to order a cutback in production of 37.5% for Occidental, 15% for several others, on the premise that the companies were producing too fast and thus harming their wells. The cut meant less oil to spread the overhead on. The Libyans demanded more taxes and told the companies to raise their own prices. . . . The Libyans knew to a very close degree just what the market would bear; the winter heating season was coming and they figured Europe could be forced to pay more." "Oil in Ferment," *Forbes,* Oct. 15, 1970, pp. 23-25. The Libyans quickly won the tax negotiations by dividing up the companies. They knew major companies like Esso, Texaco, and Mobil that operate in many other countries, would balk at granting precedent-setting increases. So they went to work on Occidental and the other independents that have relatively little oil outside Libya. The independents, with little real choice, quickly capitulated.

19. OPEC, Resolution XIII.80.

20. OPEC, Resolution XX.114.

21. Rifai, *Pricing of Crude Oil,* p. 252.

22. Ibid., pp. 258-59.

23. Rouhani, *History of OPEC,* p. 10. See also *Economist,* Nov. 1970.

24. "How the Arabs Changed the Oil Business," *Fortune,* Aug. 1971, pp. 113-16, 190-200.

25. OPEC, Resolution XXI.120.

26. OPEC, Resolution XXI.120.

27. The bulk of this information is derived from "OPEC Adopts Unified Price Strategy," *MEES*, Dec. 18, 1970, pp. 1-3.

28. Ibid., pp. 2-3.

29. Testimony of Ambassador John Irwin II in Senate, *Multinational Corporations*, pt. 5, pp. 145-50.

30. Testimony of Henry Mayer Schuler, chief London Policy Group representative for the Bunker Hunt Oil Co., in ibid., pp. 75-95.

31. For a complete text of this message, see ibid., pt. 6, pp. 60-61.

32. Rouhani, *History of OPEC*, p. 14.

33. J.E. Hartshorn, "From Tripoli to Teheran and Back," *World Today*, July 1971, pp. 291-301.

34. Ambassador Irwin's testimony in Senate, *Multinational Corporations*, pt. 5, pp. 147-49.

35. Ibid., pp. 215-16.

36. OPEC, Resolution XXII.131.

37. *Economist*, Jan. 30, 1970, p. 14.

38. Ibid., Feb. 6, 1971, p. 62.

39. "The Teheran Settlement: An Appraisal," *MEES*, Feb. 19, 1971, pp. 4-5.

40. *MEES*, Feb. 19, 1971, p. 2.

41. The following discussion is extracted from a statement by Schuler in Senate, *Multinational Corporations*, pt. 6, pp. 1-59.

42. Ibid., pp. 6-7.

43. Ibid., p. 14.

44. Ibid., p. 25.

45. Rouhani, *History of OPEC*, p. 19.

46. Schuler's statement in Senate, *Multinational Corporations*, pt. 6, p. 188.

47. Resolution XXII.131. See also Gurney Breckenfield, a petroleum researcher, who noted that "By terms of the Teheran agreement the Gulf states were bound to join Libya in an embargo only if the Tripoli Administration proved unable to win equal terms from the oil companies. Moreover, if Libya shut down its wells after making larger demands the [Teheran] agreement empowered oil companies to make up part of the deficit by faster pumping in the Gulf." "How the Arabs Changed the Oil Business," p. 200.

48. Rouhani, *History of OPEC*, p. 21.

49. Senate, *Chronology of Libyan Negotiations*, p. 15.

50. Ibid., p. 16.

51. Rouhani, *History of OPEC*, pp. 23-24.

52. Rifai, *Pricing of Crude Oil*, p. 267.

53. "How the Arabs Changed the Oil Business," p. 200.

54. Ibid.

Chapter 4. The World on a Roller Coaster

1. Peter Odell as cited by M.S. al-Mahdi, "The Pricing of Crude Oil in the International Market: Search for Equitable Criteria," Supplement to *MEES*, July 14, 1972, pp. 1-13.

2. Ibid., p. 4.

3. M.S. al-Mahdi, "Views Concerning Some Future Prospects of the Oil Exporting Countries," Supplement to *MEES*, March 30, 1973, pp. 1-14.

4. Resolution XVI.90, OPEC, *Annual Review and Record*, 1968, p. 19; Resolution XVIII.103, ibid., 1969, p. 13; and Resolution I.1 and IV.32.

5. Resolution XXI.122, OPEC, *Annual Review and Record*, 1970, pp. 24-25.

6. "The $ Crisis: Its Effect on Oil and Trade in the Middle East," *MEED*, Aug. 20, 1971, pp. 947-48.

7. "Libya Heads Off OPEC's New Offensive," *Oil and Gas Journal* (hereafter cited as *OGJ*), Nov. 1, 1971, pp. 40-41.

8. "OPEC and the Diminishing Dollar," *MEED*, Nov. 26, 1971, pp. 1360-61.

9. The "quotas," also referred to as "volume control," were based on levels of production for 1970, which was the peak year for Venezuelan crude output—over 3.7 million b/d. Under the new decree, producers were liable for heavy tax penalties if their production waivered 2 percent under or over 1970 output. The measure was taken by the government to prevent producers from using cheaper supply sources if Venezuelan prices rose above competitive levels. *OGJ*, Jan. 24, 1972, p. 40.

10. "OPEC Talks Off to Acrimonious Start," *OGJ*, Jan. 17, 1972, p. 65.

11. "OPEC Gulf States and Oil Companies Settle for 8.49 Percent Increase in Posted Prices to Compensate for Dollar Devaluation," Supplement to *MEES*, Jan. 21, 1972, pp. 1-4.

12. Jahangir Amouzegar, "The Other Side of the Oil Thing," *New York Times*, April 6, 1973.

13. "Oil Prices Formula Valid until 1975," *MEED*, June 8, 1973, pp. 643-44.

14. A further elaboration of this point was provided in the following two "hypothetical" examples: (1) In March-June 1971 just after the conclusion of the Teheran agreement with a posting of $2.18/b: production cost, 10¢; government take, $1.27; realized price, $1.70. Result: a notional company take of 33¢ and a profit split on realized prices of roughly 80/20 in the government's favor. (2) A possible present situation with posted and realized prices both around the $3.00/b mark: production cost, 10¢; government take, $1.86 (including 10¢ earnings from participation). Result: a notional company take of $1.04 and a profit split on realized prices of roughly 64/36 in the government's favor. "OPEC All Set for Price Talks," *MEES*, Sept. 21, 1973, pp. 1-4.

15. "Teheran Agreement in Need of Extensive Revision Says Yamani," Supplement to *MEES*, Sept. 7, 1973, pp. 1-3.

16. "OPEC Signals Another Crude Price Boost," *OGJ*, Sept. 24, 1973, p. 80.

17. Testimony of George T. Piercy, senior vice president and director, Exxon Corp., in Senate, *Multinational Corporations*, pt. 5, pp. 216-17.

18. For a detailed discussion of the Oct. 16 announcements, see "Unilateral 70 Percent Posted Price Increase by Gulf Producing States," *Petroleum Times*, Oct. 1973, p. 2.

19. "Gulf States Meet in Teheran on 22 December to Fix New Postings for 1 January 1974," *MEES*, Nov. 23, 1973, pp. 4-4b.

20. "Shah of Iran Explains Latest Oil Price Increase," Supplement to *MEES*, Dec. 28, 1973, pp. 1-4.

21. This estimate assumed a landed cost of $10/b. "Shock for the Third World," *Petroleum Economist*, Feb. 1974, pp. 46-48.

22. "OPEC's Latest Bombshell," *Petroleum Economist*, Jan. 1974, p. 9.

23. James Grant, "The International Petrodollar Crisis and the Developing Countries," U.S. Congress, House, Subcommittee on International Finance of the Committee on Banking and Currency, Hearings, *International Petrodollar Crisis*, 93 Cong., 2 sess. (Washington, D.C.: GPO, 1974), pp. 20-50.

24. Recycling of petrodollars serves two functions: it provides consuming states with

loans, grants, or investments and payments for goods and services; and its overall purpose is to enable those consuming nations that are unable to attract a flow of investment capital to avoid bankruptcies. Gerald Parsky, "Recycling of Oil Revenues: The Role of U.S. Capital Market," *Vital Speeches of the Day*, Feb. 25, 1975, pp. 282-84.

25. For a recent overview of Saudi oil-related motives, see William Quandt, *Saudi Arabia's Oil Policy* (Washington, D.C.: Brookings Institution, 1982).

26. For further details, see "Saudi Arabia versus the Rest at Quito OPEC Conference," *MEES*, June 21, 1974, pp. 1-5.

27. *MEED*, June 28, 1974.

28. "OPEC Raises Government Take by 3.5 Percent to Compensate for Inflation in Industrialized Countries," Supplement to *MEES*, Sept. 13, 1974, pp. 1-4.

29. "OPEC Adopts New Pricing System Based on Average Government Take of $10.12/barrel Stabilized for First Nine Months of 1975," Supplement to *MEES*, Dec. 13, 1974, pp. 1-6.

30. *Economic Report of the President*, Feb. 1975 (Washington, D.C.: GPO, 1975), p. 187.

31. *Wall Street Journal*, Feb. 10, 1975.

32. The issue of the North-South dialogue or NIEO will not be dealt with here at any length. This is a highly intricate topic which deserves separate treatment. This study will only touch upon NIEO as it affected OPEC and as the latter affected the dynamics of NIEO. For excellent and comprehensive treatments of NIEO, see Jan Tinbergen (coordinator), *Reshaping of the International Order: A Report to the Club of Rome* (New York: E.P. Dutton, 1976); and Hajo Hasenpflug, ed., *The International Economic Order: Confrontation or Cooperation between North and South* (Boulder: Westview Press, 1977).

33. *MEES*, Jan. 31, 1975, pp. 1-7.

34. For example, see the following in the *Wall Street Journal*: "OPEC Ministers Ponder Potential Crack in Solidarity Amid Falling Oil Demand," Feb. 26, 1975; "Many Analysts See Weakening of OPEC, Oppose Concession by U.S. to Oil Cartel," March 20, 1975.

35. See the following in the *Wall Street Journal*: "Saudi Arabian Output Falls Sharply," Feb. 20 and 21, 1975; and "Aramco Oil Output Falls Sharply in April, Fifth Drop in Row Reflects Slow Demand," May 7, 1975.

36. Supplement to *MEES*, March 7, 1975, pp. 1-5.

37. The preceding is based on "OPEC Summit Declaration Calls for New Deal between Industrial Nations and the Third World," Supplement to *MEES*, March 7, 1975, pp. 1-5; and Luis Turner, "The North-South Dialogue," *World Today*, March 1976, pp. 81-82.

38. "Progress Painfully Slow Toward Improved U.S. Energy Position, *OGJ*, Nov. 11, 1974, pp. 141-44.

39. Special Drawing Rights were established at the Rio de Janeiro Conference of 1967 and used for the first time in 1970. (See Glossary.)

40. When the oil companies were in charge of upstream operations, the price differentials were ironed out in the ultimate packaging of a variety of grades of crude lifted by them, even though the oil companies never tried to develop specific mechanisms to adjust for differentials. When the oil states took charge of upstream operations, however, the adjustment of differentials needed attention, since their production involved a narrow range of crudes and they lacked the operational maneuverability of the international oil companies. The issue of differentials became especially problematic under soft market conditions (which prevailed in 1974-1975). At the February meeting,

OPEC allowed its members to follow Abu Dhabi's example of reducing gravity differentials, but did not recommend any formal mechanisms.

41. *Economic Report of the President,* 1975, p. 19.

42. *The Impact of the Rise in the Price of Crude Oil in the World Economy,* Congressional Research Service Report (Washington, D.C.: GPO, 1975), p. 4.

43. The 1974 figure on current account deficit is from *Economic Report of the President,* Jan. 1976 (Washington, D.C.: GPO, 1976), p. 141.

44. See, for example, "U.S. Plans World Group of Oil Importing Nations," *New York Times,* April 16, 1973; "Casey Urges World Group of Oil Users," *New York Times,* June 22, 1973; "U.S. Pushing Front to Counter OPEC," *OGJ,* Feb. 5, 1973, p. 34.

45. For similar sentiments, see "OPEC Put Pressure on Geneva Agreement," *Petroleum Times,* March 23, 1973, p. 13.

46. "Japan Seeks Better Ties with Producing Nations," *OGJ,* Oct. 15, 1973. See also *New York Times,* April 25, 1973.

47. Romano Prodi and Alberto Clo, "Europe," in *The Oil Crisis,* ed. Raymond Vernon (New York: W.W. Norton, 1976), pp. 91-112.

48. "ECC Interest on Avoiding Confrontation with Producing Countries," *Petroleum Times,* June 1, 1973, p. 3.

49. "Japan Buys Participation Crude Oil," *MEED,* May 18, 1973, pp. 574-75.

50. From a speech delivered by Henry Simonet, EEC commissioner of energy as cited in *Petroleum Times,* Feb. 8, 1974, p. 3. U.S. Secretary of State Henry Kissinger also frequently expressed similar sentiments. For example, see his speech at the Washington energy conference, *New York Times,* Feb. 11, 1974.

51. Members were Canada, Belgium, Denmark, West Germany, Ireland, Italy, Japan, Luxembourg, the Netherlands, Norway, the United Kingdom, and the United States.

52. For a complete documentation of IEP, see "Appendix" to U.S. Congress, House, Subcommittee on International Organization and on International Resources, Food, and Energy of the Committee on International Relations, Hearing, *Legislation on the International Energy Agency,* 94 Cong., 1 sess. (Washington, D.C.: GPO, 1975), pp. 56-79.

53. Extracted from the testimony of Thomas O. Enders, assistant secretary of state for economic and business affairs, Department of State, in ibid., pp. 7-10; and U.S. Congress, House, Subcommittee on International Organization and Movements and Foreign Economic Policy of the Committee on Foreign Affairs, Hearings, *U.S. Energy Policy and the International Energy Agency,* 93 Cong., 2 sess. (Washington, D.C.: GPO, 1975).

54. C. Fred Bergsten, "The Threat from the Third World," *Foreign Policy,* Summer 1973, pp. 102-24. For a continuation of debate on this issue, see Zuhayr Mikdashi, "Collusion Could Work," pp. 56-68; Stephen Krasner, "Oil Is the Exception," pp. 68-84; and C. Fred Bergsten, "The Threat Is Real," all in *Foreign Policy,* Spring 1974.

55. "Resolutions of the Sixteenth OPEC Conference, Vienna, June 24-25," *OPEC Official Resolutions and Press Releases, 1960-1980* (New York: Pergamon Press, 1980), pp. 80-83.

56. Resolution XXIV.135, *Annual Review and Record,* 1970 (Vienna: OPEC, 1971), pp. 37-38.

57. The discussion of this meeting is based upon "OPEC Presses on with Participation Demand," *MEES,* July 16, 1971, pp. 1-4.

58. *Petroleum Intelligence Weekly,* Sept. 13, 1971, pp. 3-4.

59. This and the following discussion are based upon "OPEC Countries Plan to

Negotiate with Oil Companies on Participation and Monetary Changes," *MEES*, Sept. 24, 1971, pp. 1-6, 6a.

60. Twenty-two companies were involved in this endeavor: BP, Shell, Mobil, Gulf, Texaco, Socal, Jersey Standard, Marathon, Continental, Amerada-Hess, Occidental, Atlantic-Richfield, Grace Bunker-Hunt, CFP, Ashland, Aminoil, Sohio, Signal, Hispanoil, Petrofina, Gelsenberg, and Arabian Oil of Japan.

61. The account of this meeting is based on "Yamani Will Pursue Participation Talks with Oil Companies," *MEES*, Jan. 28, 1972, pp. 1-3; and "OPEC Spotlight Shifts to Participation," *OGJ*, Jan. 31, 1972, pp. 56-57.

62. For an insider's view on this point, see "OPEC Plans Majority Takeover If Participation Talk Fails," *MEES*, March 17, 1972, pp. 1-2.

63. "Press Release," Information Department, No. 13-72, Vienna, Nov. 30, 1972, *OPEC Official Resolutions*, p. 127.

64. "Yamani Calls for Meetings of Gulf Oil Ministers to Discuss Participation Agreement," *MEES*, Oct. 13, 1972, pp. 1-3.

65. See "Oxy's Libyan Deal Sets New Precedent in Participation Terms," *MEES*, Aug. 17, 1973, pp. 1-5.

66. "OPEC Maintains Solidarity in Face of U.S. Pressure," *MEED*, Feb. 21, 1975, p. 3.

67. M.A. Adelman, "Is the Oil Shortage Real? Oil Companies As OPEC's Tax Collectors," *Foreign Policy*, Winter 1972-1973, pp. 69-107.

68. Frank Gardner, "Competition Is Coming," *OGJ*, Jan. 13, 1975, p. 25.

69. *MEES*, March 8, 1982.

70. "Relying on OPEC to Open Crude Throttle Declared Risky," *OGJ*, Sept. 25, 1972, p. 70.

71. James E. Akins, "The Oil Crisis: This Time the Wolf Is Here," *Foreign Affairs*, April 1973, pp. 462-90.

72. "Arabs Reaffirm Use of Oil as Weapon," *OGJ*, March 24, 1975, pp. 22-23.

73. *New York Times*, Sept. 15, 1975.

Chapter 5. The Oil Embargo

1. OAPEC was formed in 1967-1968 as an exclusively Arab organization. Saudi Arabia, Kuwait, and Libya were its original members. Algeria, Abu Dhabi, Bahrain, Dubai, and Qatar joined later, and subsequently it was further expanded with the admission of Egypt, Syria, and Iraq.

2. This statement is not intended to minimize the role of the October 1973 Arab/Israeli war in bringing about an end to the impasse in the Middle East.

3. These figures are extracted from *BP Statistical Review of the World Oil Industry, 1973* (London: British Petroleum Co., 1973).

4. For details, see J.E. Hartshorn, "Oil and the Middle East War," *World Today*, April 1968, pp. 151-57.

5. No judgment is being passed concerning the legitimacy, or lack of it, of the Arab or Israeli position. This issue is complex, and a fair treatment of it is a topic in itself. For a detailed discussion of both points of view, see Edward Sheehan, "How Kissinger Did It: Step by Step Diplomacy in the Middle East," *Foreign Policy*, Spring 1976, pp. 3-70; and Yigal Allon, "Israel: The Case of Defensible Borders," *Foreign Affairs*, Oct. 1976, pp. 38-53.

6. Fuad Itayim, "Arab Oil—The Political Dimension," *Journal of Palestinian Studies,* Winter 1974, pp. 84-97.

7. A detailed explanation of the motives of OAPEC comes from Zaki Yamani in his capacity as the acting secretary general of OAPEC. He spelled out the following three motives: "(1) To keep oil activities within the organization with a view to protecting the member states from precipitous decisions and making oil a genuine weapon to serve the interests of the producing countries and Arab countries in general. (2) To create opportunities for joint investment in the oil resources of the member states. (3) To build an economic bridge between the organization and the consuming countries so as to create an outstanding economic structure which we can utilize to expand our markets and thereby establish a stronger political center of gravity for the states of the region." *MEES,* Sept. 13, 1968, pp. 6-7.

8. *BP Statistical Review,* 1973.

9. *New York Times,* April 29, 1973.

10. For example, see Senate, *Multinational Corporations,* pt. 7, pp. 504-17; *Newsweek,* Sept. 19, 1973; and interview with King Faisal on NBC, Aug. 29, 1973.

11. Itayim, "Arab Oil," pp. 88-89.

12. U.S. Congress, House, Committee on Foreign Affairs, Report of a Study Mission to the Middle East, *The United States Oil Shortage and the Arab-Israeli Conflict,* 93 Cong., 1 sess. (Washington, D.C.: GPO, 1973), pp. 90-91.

13. "Arab Oil Cutbacks Bigger Than Expected," *MEES,* Oct. 26, 1973, p. 2.

14. U.S. Congress, Senate, Subcommittee on Multinational Corporations of the Committee on Foreign Relations, report prepared by the Federal Energy Administration's Office of International Energy Affairs, *U.S. Oil Companies and the Arab Oil Embargo: The International Allocation of Constricted Supplies,* 94 Cong. (Washington, D.C.: GPO, 1975), p. 2.

15. House, *United States Oil Shortages,* pp. 15-16.

16. This discussion is based on "Arab States Raise Oil Production Cutback to 25 Percent," Supplement to *MEES,* Nov. 2, 1973, pp. 2-6.

17. Itayim, "Arab Oil," pp. 90-91.

18. "Holland Hit by Arab 'Oil Weapon,' " *MEED,* Nov. 16, 1973, pp. 1329-31, 1348-50.

19. *New York Times,* Nov. 7, 1973.

20. "Arab Summit Confirms Maintenance of Oil Weapon," *MEES,* Nov. 30, 1973, pp. 1-2.

21. One of the most obvious candidates for this increase was Britain, whose economy was suffering from the worst effects of coal miners' "go-slow" campaign to extract increased wages. For a complete background of this meeting, see "Arabs Relax Oil Production Cutbacks," *MEES,* Dec. 28, 1973, pp. 8-13.

22. The Shah of Iran, since he was not a party to the oil embargo, perceived no such threat from the West, and his drive to escalate prices of crude oil was virtually uninhibited.

23. "Worldwide Recession Is Seen As Possibility If Oil Embargo Persists," *Wall Street Journal,* Nov. 30, 1973. See also "European Unity Vital to Face the Oil Crisis," *Financial Times,* Dec. 11, 1973.

24. "Ten Percent Energy Cuts Strike Home to Japan," *Financial Times,* Nov. 19, 1973; "Japan Wakes Up to a Painful Period of Readjustment," *Financial Times,* Jan. 4, 1974.

25. "Arab Oil Restrictions May Be Eased," *MEED,* Jan. 25, 1974, pp. 108-9.

26. "Embargo on Oil Sales to U.S. Maintained," *MEED,* Feb. 22, 1974, pp. 205-6.

27. The following are some of the most frequently cited examples of leaks. The foremost source of leaks was Iraq because of its refusal to participate in general production cutbacks. Second, in the aftermath of the Nov. 4 decision of OAPEC to raise production to 25 percent, Libya and Algeria publicly announced cutback formulas different from the one adopted by the rest of the OAPEC membership. The third source was the inability of Libya to ban crude sales to Caribbean refineries, which substantially produced for the American needs, but this leak was plugged on Dec. 21.

28. For example, consider the following: "Within the industry, there was widespread belief that failure to redistribute supplies equitably would simply invite greater governmental intervention. The companies were thus anxious to avoid 'provoking' consuming governments and parliaments, making greater efforts to convince consuming nations that the 'pain was evenly spread.' On the other hand, the companies were fearful of provoking greater militancy among OPEC and OAPEC nations. This fear, principally of nationalizations or more onerous 'Participation' agreements, effectively discouraged companies from violating the embargo." Senate, *U.S. Oil Companies and the Arab Oil Embargo*, p. 5.

29. Ibid., p. 8.

30. Ibid., pp. 9-10.

31. All of the above statistics are from the Federal Energy Administration, Project Independence Report, *Project Independence* (Washington, D.C.: GPO, 1975).

Chapter 6. Toward an Uncertain Future

1. As a state with a high level of surplus production capacity, Saudi Arabia has the potential of disciplining the behavior of OPEC price hawks by substantially increasing its production level and thereby glutting the oil market, or of influencing the policies of the industrial consuming states by introducing drastic production cutbacks. In this context, it should be noted that Saudi Arabia was readily capable of such influence between 1973 and 1978. But, in view of the heightened capital needs of that country in recent years to bankroll accelerated industrialization and to purchase sophisticated military equipment, Saudi Arabia may no longer be able to introduce major reductions in its production rates with much ease.

2. *OECD Economic Outlook*, Dec. 1975, pp. 5-6.

3. U.S. crude production peaked in 1970 at the rate of 9.6 million b/d has steadily declined since then. Toward the end of 1976, it stood at 8.2 million b/d.

4. *Wall Street Journal*, March 18, 1976.

5. Larry Auldridge, "World Oil Production to Start New Uptrend in '76," *OGJ*, Feb. 16, 1976, pp. 27-30.

6. *OECD Economic Outlook*, July 1976, p. 7.

7. For a detailed background of this conference, see "OPEC Price Freeze Continues after a Stormy Session in Bali," Supplement to *MEES*, June 7, 1976, pp. 1-4.

8. Larry Auldridge, "World Oil Flow Recovers, Jumps 5% in First Half," *OGJ*, Aug. 30, 1976, pp. 19-20.

9. Walter J. Levy's study is cited in "OPEC Oil Need to Soar 4.3 Million b/d," *OGJ*, Aug. 16, 1976, pp. 52-53. Another study, conducted by Morgan Guaranty Trust and published in the Sept. issue of *World Financial Market*, labeled Levy's projections on OPEC exports "excessive."

10. *Wall Street Journal*, June 14 and Aug. 13, 1976.

11. The reason for the acutely divergent conclusions reached by the IMF and the

Economic Commission is that the former agency's indices, according to one source, are predominantly a function of inter-OECD trade, a fact that would certainly suggest the existence of a large measure of price discrimination against OPEC countries. The Economic Commission's index was based on price statistics supplied by the member governments on imported goods classified in accordance with SITC formula, had both f.o.b. data and appropriate allowance for freight, and did not include demurrage charges arising from port congestion in the OPEC countries, nor military expenditures, nor the cost of services. *MEES,* Nov. 29, 1976, pp. 1-2.

12. See "Diplomacy Used to Fight Price Hikes," p. 74, and Bart Collins, "OPEC Price Increases—Perhaps Not Inevitable," p. 75, both in *OGJ,* Nov. 22, 1976.

13. Consequently, the theoretical price of the marker crude for the eleven oil states was raised from $11.51/b to $12.70/b on Jan. 1, and it was to be increased to $13.30/b on July 1, 1977; Saudi Arabia and the UAE raised their prices to about $12.09/b. The gap between the two tiers was to remain at 61¢/b for the first half of the year and $1.21/b for the second half. See "OPEC Prices: A Parting of the Ways," Supplement to *MEES,* Dec. 20, 1976, pp. 1-7.

14. It should be noted that Yamani, on a number of occasions after the Doha conference, made it plain that his country was willing to expand its production to the maximum installed capacity of 11.8 million b/d, if necessary, to force other countries to toe the Saudi price line.

15. According to one authoritative source, the Saudis were fully aware of the extraordinarily high stock of crude and oil products in the U.S., in Japan, and, to a lesser extent, in Europe, as well as the tide of tanker cargo slow-steaming at sea. All of this added up to a glutted market, signaling reduced future liftings. The Saudis saw the handwriting on the wall and decided it was an opportune moment to do a little fence-mending with their fellow OPEC members by reunifying the two tiers of prices. "Was Need for Pressure Tests or Market Glut behind Saudi Move?" *Platt's Oilgram and News Service,* Aug. 17, 1977, p. 3.

16. In fact, it was primarily due to a substantial boost in crude production from Alaska's North Slope, which went as high as 700,000 b/d in Dec. 1977, that the U.S. output recorded an increase for that year. See Larry Auldridge, "Global Flow of Oil Climbs to High Mark during 1977," *OGJ,* Feb. 27, 1978, pp. 43-46.

17. In this context, it should be noted that the output of the U.K. rose to 850,000 b/d in Dec. while the average production for 1977 for that country remained at 800,000 b/d. Norwegian production rose to nearly 400,000 b/d in Dec. Ibid., p. 44.

18. *OECD Economic Outlook,* Dec. 1977, pp. 2-4.

19. "Saudi Arabia, Iran, UAE Determined to Continue Price Freeze into 1978 Regardless of OPEC Solidarity," *Platt's Oilgram and News Service,* Dec. 19, 1977, p. 3.

20. The above is based on "World Flow Slumps, OPEC Crude Crowded Out," *OGJ,* May 29, 1978, pp. 19-22; and "OPEC Bears Brunt of Oil Output Decline," *OGJ,* Aug. 28, 1978, pp. 35-36.

21. "OPEC Bears Brunt of Oil Output Decline," p. 35.

22. For a detailed discussion of the Taif meeting, see "OPEC: A New Direction?" *MEES,* May 15, 1978, pp. 1-4.

23. Figures on 1974 deficits and 1975 surpluses are from *OECD Economic Outlook,* Dec. 1977, p. 6; figures of 1976 surpluses and 1977 deficits are from ibid., Dec. 1978, p. vii.

24. "Oil Supplies—Deluge or Drought," *Petroleum Economist,* Oct. 1978, pp. 406-8.

25. Testimony of Arnold E. Safer, vice president, Economic Research and Planning,

Irving Trust Co., New York, in U.S. Congress, Subcommittee on Energy of the Joint Economic Committee, Hearings, *Energy in the Eighties: Can We Avoid Scarcity and Inflation?,* 95 Cong., 5 sess. (Washington, D.C.: GPO, 1978), pp. 2-20.

26. For example, see "Global Glut Called 'Dangerous Lull,' " *OGJ,* March 27, 1978, pp. 90-91; "Growing Impact on Oil Supplies," *Petroleum Economist,* March 1978, pp. 89-90; and "Trends in World Oil Demand," *Petroleum Economist,* July 1978, pp. 278-79.

27. "Implications of the Iranian Oil Supply Crisis," *MEES,* Nov. 6, 1978, p. 2.

28. The following discussion is based on the testimony of Assistant Secretary Harry Bergold in U.S. Congress, Senate, Committee on Natural Resources, Hearing, *Iran and World Supply,* 96 Cong., 1 sess. (Washington, D.C.: GPO, 1979), part 2, pp. 9-15; and also on a report by the General Accounting Office of the United States, *Analysis of the Energy and Economic Effects of the Iranian Oil Shortfall,* March 5, 1979 (Washington, D.C.: GPO, 1979).

29. In fact, as a result of such meetings in March 1979, the IEA governing board concluded that that year's oil supplies could fall short of projected oil demand by more than 2 million b/d. Consequently, the member countries resolved to stabilize the world market by reducing their collective consumption by 2 million b/d by the last quarter of 1979, a drop of about 5 percent of the projected IEA consumption. Senate, *Iran and World Oil Supply,* part 4, p. 3.

30. For further details of this conference, see "New Oil Prices Please Most of OPEC," *OGJ,* Dec. 25, 1978; and "OPEC: 1979 Prices Settled But Future Remains Cloudy," *MEES,* Dec. 25, 1978, pp. 1-6.

31. According to one report, the spot market may have accounted for about 25 percent of world crude oil trade in 1979, and American imports from that market averaged about 9 percent of all U.S. crude oil imports between April 1979 and February 1980. The significant role of spot markets in this period also becomes apparent by the fact that in 1979 spot-priced crude originated in twenty-three of twenty-seven OPEC and non-OPEC countries from which the United States imported crude oil. Report by General Accounting Office of the United States, *The United States Exerts Limited Influence on the International Crude Oil Spot Market,* Aug. 21, 1981, pp. i-ii (Washington, D.C.: GPO, 1981).

32. The account of this meeting is based on "A Multi-tier Price Structure for OPEC," *MEES,* April 2, 1979, pp. 1-4; and "OPEC Accelerates Price Hike Schedule," *OGJ,* April 2, 1979, pp. 44-45.

33. It was clear, however, that by allowing its members to determine the amount of market premiums or surcharges, OPEC was at least indirectly responsible for a free-for-all on surcharges. For further details, see the preceding footnote.

34. The preceding discussion is extracted from GAO, *U.S. Exerts Limited Influence,* pp. 3-4.

35. For a detailed discussion of havoc created by spot prices, see *MEES,* April 9, May 7 and 21, 1979; also "More Surcharges Hike OPEC Oil Prices," *OGJ,* May 28, 1979, pp. 56-58. For a complete discussion of oil policy in post-revolutionary Iran, see Shaul Bakhash, *The Politics of Oil and Revolution in Iran* (Washington, D.C.: Brookings Institution, 1982).

36. See *OGJ,* July 2, 1979; *MEES,* July 2, 1979, pp. 1-4.

37. For example, see "OPEC Price Ceiling in Jeopardy," *MEES,* Sept. 17, 1979, pp. 1-2; "Kuwait Raises Prices: Insists on Traditional Differentials," *MEES,* Oct. 15, 1979, pp. 1-3; and "Bang Goes the OPEC $23.50/Barrel Ceiling," *MEES,* Oct. 22, 1979, pp. 1-2.

38. The financial implications of such practices become abundantly clear in the following estimates: "Assume an average price of $3 a barrel and multiply that by 6.5 million barrels a day that the Aramco partners have been allowed to take recently. The result is possible additional profits, before taxes, of $19 million a day, or about $7 billion annually." "Four U.S. Oil Concerns Benefit as Saudis' Prices Trail Rest of OPEC's," *Wall Street Journal,* July 24, 1979. The article reports the total 1978 net income of the four Aramco companies to be $584 billion.

39. Following are percentage gains made by other oil companies in the period under discussion: Sohio posted a 191 percent gain; Conoco, 154 percent; Marathon, 58 percent; Standard Oil of Indiana, 49 percent; and Atlantic Richfield, 45 percent. *Wall Street Journal,* Oct. 25, 1979.

40. *Facts on File,* 1980, p. 90.

41. For example, consider the following statement by Yamani: "The extra profits made by the oil companies in the U.S. are, unfortunately coming from Saudi Arabia. They are filling their refineries with cheap Saudi crude . . . and reselling it in the form of products, whose prices reflect the higher cost of crude at spot market prices or the official prices of other OPEC countries." *MEES,* Nov. 5, 1979, p. 1.

42. According to one estimate, the cumulative impact of such production cutbacks, including the possible Saudi Arabian measure to lower its own production by 1 million b/d, was expected to result in a total cut of more than 3 million b/d in an already tight market. See Roger Vielovye, "Confusion Clouds OPEC Caracas Meeting," *OGJ,* Dec. 17, 1979, p. 44.

43. *Wall Street Journal,* Sept. 24, 1979.

44. *OECD Economic Outlook,* July 1980, pp. 3-5.

45. *OECD Economic Outlook,* Dec. 1980.

46. "Global Oil Swings from Shortage toward Surplus," *OGJ,* Feb. 25, 1980, pp. 27-31.

47. "OPEC Members Settling on Crude Prices," *OGJ,* Jan. 7, 1980, pp. 39-40.

48. *OPEC Chronology* (Vienna: OPEC, 1980), pp. 63-64. Further evidence of the continued inability of the Saudis to reunify OPEC prices on their terms, i.e., by lowering them, was apparent in Yamani's response to a question as to whether Saudi Arabia would stay at the $28/b scale. "We will watch the market and we will act accordingly. Now I won't be surprised if I stay at $28.00 as I am, probably until we meet next September. But I don't rule out the possibility of an increase, not necessarily to $32.00; it could be one dollar or so. So the whole situation is not yet clear for Saudi Arabia; we will watch the market forces and see what will happen." "OPEC: A Tentative Step toward Price Reunification," *MEES,* June 16, 1980, pp. 1-7.

49. The discussion of this meeting is extracted from "OPEC Sets Base Marker Crude Price at $30/Barrel," *MEES,* Sept. 22, 1980.

50. Prior to the outbreak of hostilities, Iran was reported to be producing about 1.4 million b/d, and Iraqi output remained a little over 3 million b/d. As a result of war-related damages, Iran stopped its exports while Iraq was reported to be exporting between 800,000 and 1 million b/d through pipelines to Mediterranean ports, and about 400,000 to 500,000 b/d was reported to be moving through the Iraqi-Turkish pipeline. Also see Table A-2.

51. See testimony of Charles L. Campbell, senior vice president, Gulf Trading and Transportation Co., in U.S. Congress, House, Subcommittee of the Committee on Government Operations, Hearing, *Effect of Iraqi-Iranian Conflict on U.S. Energy Policy,* 96 Cong., 2 sess. (Washington, D.C.: GPO, 1981), pp. 42-43.

52. *MEES,* Oct. 6, 1980, p. 4.

53. Even though the IEA's emergency sharing plan was created to deal with a potential situation described here, the fact that the effectiveness of such a program did not undergo the rigorous testing of a real shortage became a real source of concern among OECD members every time they encountered a potential crisis affecting oil supply.

54. Saudi production was raised from 9.5 million to 10.3 million b/d; Kuwaiti, from 1.3 million to 1.5 million b/d (although this decision was made by Kuwait prior to the Iraq-Iran war and no additional increases were considered in response to the war); the UAE raised its production by about 100,000 b/d, and Qatar, by 20,000 b/d.

55. For a detailed discussion of Adelman's scenarios, see Chapter 4, above.

56. For complete details, see Robert Mabro, "The Changing Nature of the Oil Market and OPEC Policies," *MEES,* Sept. 20, 1982, pp. 1-9.

57. This reduction applied to about 550,000 to 600,000 b/d of crude which Saudi Arabia was producing from Oct. 1980 on for consumers affected by the Iran-Iraq war. The bulk of Saudi crude was being sold at $32/b.

58. "Low Liftings Hold Lid on OPEC Prices," *OGJ,* July 20, 1981, pp. 17-21; "OPEC Struggling to Unify Crude Oil Prices," *OGJ,* Aug. 24, 1981, p. 53.

59. Robert J. Enright, "World Oil Flow, Refining Capacity Down Sharply; Reserves Increase," *OGJ,* Dec. 27, 1982, pp. 75-82.

60. The preceding data are extracted from Robert J. Beck, "U.S. Demand to Fall Again by 4.2%; Imports Also Slide, But Production Up," *OGJ,* July 26, 1982, pp. 175-204.

61. Enright, "World Oil Flow," p. 76.

62. *OECD Economic Outlook,* Dec. 1982, pp. 7-17, *passim.*

63. Ibid., July 1983, pp. 25-26. The forecasts of recovery in OECD countries issued in July and Dec. 1982 were gloomier.

64. The preceding statistics are extracted from "Long Decline in World Oil Output Begins to Level Off," *OGJ,* Sept. 12, 1983, pp. 57-60.

65. For further details, see "OPEC Buckles Down to Long Term Strategy," *MEES,* Nov. 21, 1983, pp. A1-A4, A14.

66. "EIA: Oil Prices Won't Rebound Quickly," *OGJ,* June 4, 1984, pp. 48-49.

67. "World Oil Situation," *1984 World Economic Outlook,* pp. 127-36.

68. The futures markets emerged in the early 1980s when the prices of crude oil were fluctuating widely. In the futures markets sellers and buyers "hedge" and "speculate" on prices in order to minimize their risk in the face of price fluctuations. The various futures markets include the New York Merchandise Exchange (NYMEX), International Petroleum Exchange (IPE), Chicago Board of Trade (CBT), and Chicago Merchandise Exchange (CME). For a comprehensive study of the futures markets, see Hussain Ravazi, *Oil, Futures Trading: The Impact on the Structure of the Petroleum Market* (London: Financial Times Business Information, 1984).

69. The netback for Arabian Light refined in the Mediterranean, after rising to $27.70/b, dropped back to $27.10/b. The spot price of U.K.'s Brent crude fell by $3 to just over $26/b before partially recovering. Norway was encountering difficulty at selling its crude at $29/b and believed its crude should sell at $28/b or possibly lower. See *Petroleum Economist,* Nov. 1984, p. 431.

70. Frank Neiering, Jr., "Market Trends," *Petroleum Economist,* Oct. 1984, pp. 395-96.

71. *Petroleum Economist,* Nov. 1984, pp. 402, 424, 431.

72. *OGJ,* Oct. 29, 1984, p. 38.

73. Mexico announced a reduction of at least 100,000 b/d and Egypt reduced its production by 30,000 b/d. "OGJ Newsletter," *OGJ,* Nov. 12, 1984.

74. Roger Vielvoye, "Outlook Fuzzy for World Oil Prices during 1985," *OGJ,* Dec. 31, 1984, pp. 41-44.

75. "Market Rejects OPEC Price Efforts," *OGJ,* Jan. 7, 1985, pp. 62-64.

76. Ibid., p. 64. According to another source, BNOC was reported to be selling as much as 80,000 b/d at spot market prices. The same source also indicated that BNOC planned to begin setting its prices "on the average price of the previous month's spot market transactions." Youssef M. Ibrahim, "Britain Confirms Sales of Crude Oil at Market Prices," *Wall Street Journal,* Jan. 9, 1985.

77. Youssef Ibrahim, "Norway, Britain Quietly Moving Closer to Linking Oil Prices to the Free Market," *Wall Street Journal,* Jan. 15, 1985.

78. This and the following discussion are based on Roger Vielvoye, "OPEC will Tackle Differentials Again," *OGJ,* Jan. 14, 1985, pp. 37-38.

79. Youssef Ibrahim, "British Abandon Official Pricing for Oil," *Wall Street Journal,* Jan. 8, 1985.

80. "OPEC Prices in Perspective," *Petroleum Economist,* March 1985, pp. 78-80. Also see Paul Hempt, "Dutch Accountants Take on a Formidable Task: Ferreting Out 'Cheaters' in the Ranks of OPEC," *Wall Street Journal,* Feb. 26, 1985.

81. "OGJ Newsletter," *OGJ,* Feb. 18, 1985.

82. BNOC was established in the mid-1970s to give the government control over Britain's North Sea oil production, to assure adequate oil supplies, and to set official prices. To protect British supplies, BNOC entered into an agreement to buy 51 percent of North Sea production from the participating companies at the official price and to resell the oil at the same price. In the chaotic market of the 1980s, however, BNOC had been losing money because it was buying oil at the official price but, due to lack of storage facilities, was forced to sell it at lower spot prices.

83. For a detailed discussion of the threatening posture of the Khomeini regime toward the Persian Gulf region, see M.E. Ahrari, "Implications of the Iranian Political Change for the Arab World," *Middle East Review* 16, No. 3 (Spring 1984): 17-29.

84. For details, see U.S. Congress, House, Report Prepared by the Foreign Affairs and National Defense Division of the Congressional Research Service of the Library of Congress for the Subcommittee on Europe and the Middle East of the Committee on Foreign Affairs, *Saudi Arabia and the United States: The New Context in an Evolving "Special Relationship,"* 97 Cong., 1 sess. (Washington, D.C.: GPO, 1981).

85. Throughout this listing, a "free-for-all" attitude of oil states concerning differentials and surcharges is not taken into consideration. The differential situation remained so chaotic in 1979 and 1980 that a systematic account is virtually impossible.

Chapter 7. Epilogue

1. Even prior the 1960s, all the way back to the colonial era, the relationship between the oil companies and the oil states was perceived by the latter as exploitive and asymmetrical. For details, see Chapter 1.

2. According to Murray Edelman, "Condensation symbols evoke the emotions associated with the situation. They condense into one symbolic event, sign, or act patriotic pride, anxieties, remembrances of past glories or humiliations, promises of future greatness: some of these or all of them." *The Symbolic Uses of Politics* (Champaign-Urbana: Univ. of Illinois Press, 1972), p. 6.

3. Bob Beck and Mike Obel, "World Crude, Condensate Flow in '84 Logs First Increase since 1979," *OGJ,* March 11, 1985, pp. 41-44.

4. The figure 32.1 percent is from ibid., p. 44. At the beginning of 1984, estimated commercial stocks on land in the OECD area were about 69 days of forward consumption, compared with about 75 days a year earlier. Even though this was a considerable decrease when compared to the estimated stock level of more than 95 days of forward consumption at the end of 1980, the industrial consumers were well insulated against any supply exigency because of the oil companies' improved capability to manage stocks and refineries. *World Economic Outlook, 1984,* p. 136.

5. Fereidun Fesharaki and David T. Isaak, *OPEC, the Gulf and the World Petroleum Market: A Study in Government Policy and Downstream Operations* (Boulder: Westview Press, 1983), p. 130.

Index